RAPID METHODS IN CLINICAL MICROBIOLOGY

Present Status and Future Trends

ADVANCES IN EXPERIMENTAL MEDICINE AND BIOLOGY

Recent Volumes in this Series

RAPID METHODS IN CLINICAL MICROBIOLOGY

Present Status and Future Trends

Edited by

Bruce Kleger

Pennsylvania Department of Health
Lionville, Pennsylvania

Donald Jungkind

Thomas Jefferson University
Philadelphia, Pennsylvania

Eileen Hinks

United Hospitals, Inc.
Cheltenham, Pennsylvania

and

Linda A. Miller

Holy Redeemer Hospital and Medical Center
Meadowbrook, Pennsylvania

PLENUM PRESS • NEW YORK AND LONDON

Library of Congress Cataloging-in-Publication Data

Eastern Pennsylvania Branch of the American Society for Microbiology
 Symposium on Rapid Methods in Clinical Microbiology (1987
 Philadelphia, Pa.)
   Rapid methods in clinical microbiology : present status and future
 trends / edited by Bruce Kleger ... [et al.].
      p.   cm. -- (Advances in experimental medicine and biology ; v.
 263)
   "Based on the proceedings of the Eastern Pennsylvania Branch of
 the American Society for Microbiology Symposium on Rapid Methods in
 Clinical Microbiology, held November 12-13, 1987, in Philadelphia,
 Pennsylvania"--T.p. verso.
   Includes bibliographical references.
   ISBN-13:978-1-4612-7886-3   e-ISBN-13:978-1-4613-0601-6
   DOI: 10.1007/978-1-4613-0601-6

   1. Diagnostic microbiology--Technique--Congresses.   I. Kleger,
 Bruce.   II. American Society of Microbiology.   Eastern Pennsylvania
 Branch.   III. Title.   IV. Series.
   [DNLM: 1. Microbiological Technics--trends--congresses.   W1 AD559
 v. 263 / QW 25 E13r 1987]
 QR67.E27   1987
 616.07'5--dc20
 DNLM/DLC
 for Library of Congress   90-6776
                                                                CIP

Based on the proceedings of the Eastern Pennsylvania
Branch of the American Society for Microbiology Symposium
on Rapid Methods in Clinical Microbiology, held November 12–13,
1987, in Philadelphia, Pennsylvania

© 1989 Plenum Press, New York
Softcover reprint of the hardcover 1st edition 1989

A Division of Plenum Publishing Corporation
233 Spring Street, New York, N.Y. 10013

ORGANIZING COMMITTEE

CHAIRMAN

Bruce Kleger, Dr.P.H.
Pennsylvania Department of Health

CO-CHAIRMEN

Donald Jungkind, Ph.D.
Thomas Jefferson University Hospital
Eileen Hinks, Ph.D.
United Hospitals, Inc.

COMMITTEE MEMBERS

Carl Abramson, Ph.D.
Pennsylvania College of Podiatric Medicine
Josephine Bartola, J.D.
Pennsylvania Department of Health
Walter Ceglowski, Ph.D.
Temple University School of Medicine
Kenneth R. Cundy, Ph.D.
Temple University School of Medicine
Alan T. Evangelista, Ph.D.
Cooper Hospital/University Medical Center
Gary Haller, Ph.D.
SmithKline Bioscience
Victor Iacocca, Ph.D.
SmithKline Diagnostics
Linda A. Miller, Ph.D.
Holy Redeemer Hospital and Medical Center
Irving Nachamkin, Dr.P.H.
Hospital of the University of Pennsylvania
Ron Neimeister, M.P.H.
Pennsylvania Department of Health
Joseph F. Pagano, Ph.D.
SmithKline Diagnostics
James E. Prier, Ph.D.
Philadelphia College of Osteopathic Medicine

EX-OFFICIO MEMBER

James A. Poupard, Ph.D.
President, Eastern Pennsylvania Branch, ASM
Medical College of Pennsylvania

PREFACE

The papers published herein comprise the presentations given at the eighteenth of an annual series of clinical symposia arranged under the auspices of the Eastern Pennsylvania Branch of the American Society for Microbiology. This symposium allowed approximately 200 persons to gather and exchange ideas on the rapid laboratory diagnosis of infectious diseases.

The institution of the Diagnosis Related Group (DRG) method for reimbursement by both government agencies and private insurance carriers has provided a financial aspect to the established clinical reasons for rapid laboratory diagnosis. Now the health of the institution, as well as the patient, is dependent on a timely diagnosis and, hopefully, cure. Accordingly, the goal of this symposium was to present the latest developments in "same-day microbiology".

In the face of stable or diminishing resources, the laboratory director is presented with many choices. Do nucleic acid probes, non-instrumental ELISA techniques, or time-resolved fluorometry have a place in his or her laboratory? Should the laboratory test for newly described human pathogens such as human immunodeficiency virus or human papilloma virus? Can rapid techniques supplant conventional methods? Or are they merely adjunctive? This symposium attempted to assist in the formulation of informed decisions.

Bruce Kleger
Donald Jungkind
Eileen Hinks
Linda A. Miller

ACKNOWLEDGEMENTS

We would like to thank the Eastern Pennsylvania Branch of the American Society for Microbiology for sponsoring this symposium and for making this publication possible. We especially thank the Symposium Committee for their diligent work in organizing an informative and successful symposium.

The editors acknowledge the support and sponsorship of the Bureau of Laboratories of the Pennsylvania Department of Health, Temple University School of Medicine, Hahnemann University, The Medical College of Pennsylvania, the University of Pennsylvania School of Medicine, Thomas Jefferson University and the Philadelphia College of Osteopathic Medicine.

This symposium would not have been possible without the financial support of the following companies: Abbott Laboratories Diagnostics Division, DuPont Medical Products, Gen-Probe, Marion Laboratories, Inc., Merck, Sharp & Dohme, SmithKline Diagnostics, Inc., E.R. Squibb & Sons, Whittaker M.A. Bioproducts, Inc. We are grateful for their contributions.

We would especially like to thank Josephine Bartola and the Pennsylvania Department of Health-Bureau of Laboratories for generously providing expertise and assistance in mailing and registrations. Thanks also to Donna May for her assistance in the preparation of some of the camera ready copies for this publication.

CONTENTS

"WHERE WE'VE BEEN AND WHERE WE MAY BE GOING"

Henry D. Isenberg

Microbiology Division
Long Island Jewish Medical Center
New Hyde Park, NY 11042
Department of Pathology
Health Science Center
State University of New York
Stony Brook, NY 11794

Systematic microbiology is a recently developed discipline. Microbiologists have always been interested in what microorganisms do rather than
what microorganisms are. We have been concerned with the impact of microorganisms on human and animal health, on our food supply and the integrity
of manufactured goods, but have neglected the microbial world and its
functions. As Stanier has pointed out, the process that unifies the study
of the diverse forms called microorganisms is the pure culture technique.
This approach addresses all acellular organisms be they prokaryotes or eukaryotes, despite their qualitatively different expressions of the living
process and despite the varying complexity with which they use the same
select essential compounds, fundamental pathways and mechanisms of organization.

The science of microbiology advanced in several stages. We must start
our view of where we have been by remembering the observations of
Fracastorius (1546) who thought that infection might be due to a contagium
vivium, followed by the discoveries of van Leeuvenhoek (1683) who, through
his great skill as a lens maker, was able to observe what were obviously
microbial forms in a large variety of materials from human beings, animals
and plants. Perhaps these striking observations impressed von Plenciz (1762)
who felt strongly that microorganisms cause disease in humans (14). These
intimations of a microbial world and its potential role in causing diseases
were strongly influenced by efforts to understand the process of fermentation. While J. Liebig maintained that fermentation was a chemical rear-
rangement which arose as a result of conditions within the ferment,
Cagniard-LaTour (1836), Schwann (1837) and von Helmholtz (1843) claimed that
ethanol was produced by yeast. We must also remember that the controversy
of spontaneous generation influenced subsequent perceptions of microorganisms
greatly. Needham (1745) showed that heating a mixture of organic material
did not prevent putrefaction. In 1769, Spalanzani demonstrated that increasing the heat of these solutions would result in sterility, followed in
the nineteenth century by the work of Appert (1810), Schwann (1837) and
von Helmholtz (1843) who indicated that heating the air in contact with
these boiled solutions resulted in the absence of growth activities, dis-
cernible grossly. Schultze sterilized the air by passing it through con-

1

centrated potassium hydroxide indicating once more that the subsequent putrefaction could be prevented by eliminating some vital force which entered
with air or oxygen. Schroeder and Dusch (1854) passed air through cotton
plugs and thereby prevented the entry of what we now know as microorganisms
into various organic mixes and suggested indeed that some microbial life may
gain access in this manner.

   All these hints of the possibility that microorganisms might influence
vital activities were clarified through the great contributions of Louis
Pasteur (1822-1895). This inventive genius single-handedly introduced ster-
ilization showing that steam, steam under pressure, namely, the autoclave
and hot air ovens, i.e. temperatures in excess of 180°C, in conjunction with
the use of cotton plugs, resulted in sterile solutions. Similarly, his
studies indicated that many aspects of microbial nutrition are varied and
that by the addition of certain particular constituents a selective liquid
medium could be provided. He also demonstrated clearly the influence of
pH and of the partial pressure of oxygen on the activities of microorganisms.
Pasteur, by studying fermentation and putrefaction, was able to demonstrate
the microbial origin of these important processes, a discovery that led
Lister to institute aseptic technique in the surgical theaters of England.
Pasteur demonstrated that by continued growth in the laboratory, the harmfulness of microorganisms could be attenuated by several techniques, making
them useful as vaccines. He went further and demonstrated that pure cultures of organisms obtained by serial subcultures of broth dilutions were
capable of reproducing disease. In addition, and without any of the modern
tools, he suggested that animal passage of viruses was possible. Pasteur,
the chemist, provided a large number of ideas, concepts and theories for
the embryonal science of microbiology, concepts that are still of great
moment in our present perception of microbiology. He combined his hypoth-
eses with simple elegant experiments which allowed him to place microbiology
on firm footing as an applied science for solving problems in human health,
industry, agriculture and animal husbandry.

   Robert Koch (1843-1910), on the other hand, approached microbiology in
a systematic manner. His first contribution was the life cycle of Bacillus
anthracis, a major advance in understanding various bacterial functions.
His exposure to botany led him to apply aniline dyes to visualize microorganisms better. At the suggestion of one of his colleagues (Brefeld), he
added gelatin to media in order to solidify the substrate and to isolate
microorganisms for the study of the pure cultures, an approach successful
with gelatinase-negative microorganisms. The addition of agar at the suggestion of Fanny Hesse provided Koch and his disciples with a means of investigating isolated microorganisms and obtaining pure cultures on solid
media, one organism separated from all others. An explosion of discoveries
followed this initial contribution of Koch's with his isolations of Mycobacterium tuberculosis and Vibrio cholerae. Many of his coworkers and
colleagues discovered most of the bacteria involved in infectious disease.
Koch's emphasis on technique led him to enunciate his famous postulates
which attempted to define the conditions that led to the expression of
microbial harmfulness, i.e. pathogenicity in the host. Koch differed from
Pasteur by emphasizing strict adherence to technique and efforts to define
the role for microorganisms in disease production.

   While Koch and his group were mostly concerned with the etiological
agents of human and animal disease, other microbiologists realized the need
to study the oxidation of ammonia in soil and the ability of microorganisms
to fix nitrogen. Other workers began to explore environmental microbiology
as well as plant pathology (14).

   This haphazard growth of microbiology with emphasis on the diagnosis
of infectious disease generated a number of efforts to elucidate metabolic

2

and biosynthetic pathways, chemotherapy against infectious disease and neo-plasia, molecular biology and the great discipline of immunology. While the extant practices of this last specialty encompass more than concern of the host's responses to infectious agents, the host-parasite interactions and the disturbance of this equilibrium by disease and/or therapy has ac-quired increased significance. Our understanding of infectious disease has been dominated by Koch's postulates, an alleged definition of pathogenicity an by implication, virulence. This approach remains the major concept gov-erning our attitudes toward infectious disease despite our present recog-nition of polymicrobic and hospital-acquired infections in addition to the discoveries of new bacterial and virus-associated diseases. These develop-ments indicate clearly that the individual host plays a determinant role in the overt clinical manifestations of infectious disease and that the initi-ation of an infectious process is not the mere entry of a certain agent into the human biosphere. The recent changes in our perceptions of infectious disease require a review of Koch's influence with regard to infectious disease. Above all, we must remember that the explanations of Koch succeeded in a very large number of instances. The adherence to the postulates is, therefore, understandable. The considerable influence that these attitudes exert on clinical and diagnostic microbiology require a brief review of Koch's postulates (14). Koch stated that an organism should be found in all patients with a disease in question; that the organism should be cultivated outside the body of the host in pure culture for several generations and when introduced into a laboratory animal lead to the disease. This general description of what might best be termed infectivity ignores several sig-nificant considerations. Not all organisms that produce disease can be isolated in the laboratory. Some require special stains for detection of their presence in host tissues. The postulates ignore the portal-of-entry and the minimal infective number, variables of considerable significance in clinically noticeable disease. We cannot ignore that not all non-immune hosts succumb to lethal challenges with the organism in the laboratory. The mere presence or isolation of any microorganism is not necessarily fol-lowed by disease manifestations. As mentioned already, the presence of polymicrobic infections and of hospital-acquired infectious disease indicate clearly the shortcomings of Koch's postulates (6).

Theobald Smith (12) anticipated these later developments when he pro-posed consideration of the host-parasite equilibrium and enunciated the concept of parasitism as "a universal biological process that evolved in the predatory struggle for food and, therefore, represents the normal inter-dependence of all living forms." These considerations have a profound effect on the role of the microbial world, infectious disease and its diagnosis in the laboratory. The concept of parasitism insists that the entry of a micro-organism into a particular host is left mostly to chance. This chance encounter results in the expression of microbial attributes which may be injurious to the host but which serve an entirely different purpose in the organism's original environment. An organism, therefore, might be harmful to one type of host and harmless to another. The difference may be species determined or may be intraspecies in its manifestations. This consideration of a parasite reinforces the well known observation that a microorganism can be harmful in one anatomic locus and totally harmless in another site. Obviously, there may be variations within a microbial species with regard to its ability to damage a host just as there may be variation in the host's susceptibility to any parasite's harmful activities. Certainly Smith's hypothesis establishes a role for the resident microbiota in preventing or encouraging colonization or entry by a newcomer. It further strengthens our perception of how microorganisms find a favorable climate within the intimate environment and underlines the possibility that specific receptors on certain tissues may constitute a favorable area for microbial prolifer-ation. Possibilities inherent in Theobald Smith's approach produce an altered appreciation of infectious disease by underlining the selectivity

3

exercised by the host, the involvement of tissues and the capacity of tissues and the capacity of a microorganism to destroy or damage cells leading to generalized disease. With this in mind, we must view infectious disease as a progression of interactions between the microorganism and the host with both acting in concert for the overt expression of harmfulness.

The numerous factors that bear on these considerations are outside the realm of this topic.

Our understanding of the events that permit any microorganism to settle into our intimate biosphere has been advanced considerably and is of great moment in our present practices of clinical microbiology. Briefly, any organism in our environment must be equipped to attach itself and succeed in obtaining adequate nutrients as part of the microbial consortium that populates our skin and mocous membranes. Attachment mechanisms for bacteria vary in accordance with their gram characteristics. They may consist of fimbriae, sometimes designated as pili - a term now reserved for quasisexual bacterial unions - and a variety of exopolysaccharides that form more or less organized cements. At their outermost levels these cements, produced in varying degrees by individual bacteria, interact to form a secure, albeit dynamic, microbial cover on all exposed body surfaces. The gram negative bacteria, thus, have the means to overcome the repulsion inherent in all biological structures and the means to resist the turbulence that prevails in all biological microenvironments. While fimbrial structures have been noted in some gram positive organisms, the majority project teichoic or lipoteichoic acids through their cell walls which, in concert with some extracellular proteins, react with appropriate mammalian membrane constituents to anchor these bacteria. Attachment is secured by the production of exopolysaccharide cements or gums. Similar mechanisms permit organism to organism contact resulting very often in a multimicrobial population on body surfaces, the conditions we describe as colonization. Regardless of injury or benefit derived by the host from the presence of this microbiota, we must remember that laboratory analyses oppose this natural state. We separate each bacterium from members of the same species and all other microbial forms in any specimen, provide a very rich diet and encourage the fastest growing variants to outnumber all others very rapidly. This action selects organisms that shall not waste efforts to produce attachment structures or cements. We have succeeded in isolating the laboratory variants that often lack outermost surface structures presented to the host's immune system.

Another aspect also plays a role in our perception with respect to potentially harmful bacteria. In nature, the exopolysaccharide cements of the colonizing organisms do not respond easily to mammalian enzymes. They are obstacles ignored by host immune systems. The microbial members of our colonizing microbiota are, thus, outside the ken of our immune system. Those bacteria that leave the safety of the consortium and achieve occasional entry beyond the confines of our outermost epithelia present a very different array of haptens and antigens to our immune system. The host-immune response is directed against structures that differ qualitatively and quantitatively from those of colonizing organisms. If an organism in tissue or in the circulation can find an appropriate niche for attachment and cement production, it escapes once more the policing activities of the immune system and has the potential to form a nidus for chronic infections. Many assorted host factors ranging from appropriate surface structures to the entire spectrum of immunity control the success of the harmful activities expressed by bacteria in the usual sterile parts of the host. This emphasis on the determinant role played by the host extends to the microbial eukaryotes and viruses, although less is known about the details of these interactions (6).

4

Successful isolation of any microorganism must be followed by identification (9). Parenthetically, this essay shall not concern itself with viruses and their laboratory manipulation in deference to those who practice this specialty exclusively. Traditional approaches have dominated this effort until very recently. Automated and manual systems modalities are available, but most are dedicated to identification using time-honored carbohydrate substrates and a few microbial enzymes. While yeast identification reflects this dedication of microbiologists to assess the biochemical versatility in manipulating sugars, most filamentous fungi and protozoa are identified morphologically. One cannot deny that the efforts to simplify this test by manual or automated approaches have reduced the inordinate labor demands and test variations of yesteryear; the use of substrates that produce rapid results have diminished the time required for identification and systematized the identification of especially bacteria comparatively rarely encountered in the intimate human biosphere. The application of molecular techniques to bacterial taxonomy, however, emphasizes that the 30 or even 50 phenotypic characters we can now examine represent but a small percentage of the genetic armamentarium of each species. Even DNA hybridization studies have not yet reached the ultimate goal of bacterial identification. The methodology cannot recognize important intraspecies variants. For example, it fails to distinguish the shigellae from Escherichia coli. While the establishment of a place for the shigellae as members of E. coli confirms the suspicions of earlier taxonomists, it would be clinically unacceptable to be taxonomically correct. Fortunately, the workers in molecular taxonomy appreciate the clinical microbiologists' dilemma and have not advocated the use of newly designated species or subspecies until appropriate laboratory tests make their clinical laboratory identification possible.

Our appreciation of these taxonomic advances must be tempered, however, by several circumstances with considerable impact on the practice of clinical microbiology. An important aspect of our discipline is its clinical relevance. While the need to provide important relevant information to clinicians as soon as possible has been advocated (7,13), the clinical relevance of identifications that delineate all organisms in polymicrobic specimens has been questioned by several experts (1,10,11), especially in the present climate of fiscal restraints. We must also appreciate that certain bacteria such as Escherichia coli, Klebsiella pneumoniae and Proteus mirabilis occur with great frequency in clinical specimens (2). These factors must be considered as clinical microbiology continues to develop in an effort to attain diagnostic rather than confirmatory status.

However, this understandable effort to focus microbiological analysis of clinical specimens on the most relevant information must consider the responsibility clinical microbiologists assume with respect to hospitalized patients, employees of medical facilities and the community at large. While often ignored or relegated to a minor role in publications on epidemiology, the clinical microbiology laboratory is a significant if not pivotal information center for the recognition of those organisms that must be prevented from complicating the recovery of other patients in the institution or that could be harmful to the staff or community residents. Unfortunately, there is no ready guide to help clinical microbiologists to escape this dilemma in the present setting of continuing dependence on the pure culture dogma.

Guidance for anti-infective therapy is a major effort in most clinical microbiology laboratories. Our present approaches are also not sufficiently rapid to play the decisive role one would expect. The need to isolate the potential organism of causal relationship to the observed clinical symptoms and the plethora of available agents delay useful guidance sufficiently to force clinicians to initiate therapy in the gravely ill patient without the benefit of specific information. When the laboratory report finally becomes available, it lacks the momentum of more expeditious results. At best, some

5

physicians change therapy in keeping with reports that indicate the inappropriate nature of the empirically chosen drug (8). Obviously the stumbling block to expeditious reports remains our dependence on the pure culture technique. Also, we must keep in mind the aforementioned unnatural circumstances that attend our analyses, especially our inability to mimic events occurring in the natural setting.

My exposure of these shortcomings of microbiology is not an indictment of our efforts. It is the consequence of the nature of the microbial world that so far has denied us the capability to examine a single microorganism and forces us to study isolated living populations. Clinical microbiology differs from other clinical laboratory disciplines because it must sequester pertinent information from a very dynamic conglomerate of living organisms. In order to change clinical microbiology from a confirmatory to a diagnostic specialty, our future goals must be directed at the recognition of etiological agents in the specimens we receive and the demonstration of microbial constituents or enzymes to direct appropriate therapy.

Advances in molecular and immunological techniques bring us much closer to realizing this old ambition. A cursory examination of the evolving methodology reveals almost unlimited applications to our efforts. Theoretically, our problems appear to be solved and we can join our colleagues in providing immediate information of moment to the clinicians. Unfortunately, there are several stumbling blocks that must be circumvented before this nirvana can be attained. Our collective ignorance about the normal microbiota presents the first obstacle, amplified by the present inability of molecular analysis to differentiate intraspecies variation. We must admit that our skills in demonstrating microorganisms have not allowed complete assessment of the entire spectrum of the microbial residents in or on our bodies, possible sources of confusion when molecular probes are employed. Many instances of overt infectious disease have obscure beginnings; we must ask: is an individual species capable of initiating disease; how many microorganisms or viruses must be present initially to produce the symptoms in the natural setting; are all or most infections monomicrobic or polymicrobic; is the overt expression of an infection the result of a sequence of microbial activities that involves various and sundry representatives of the microbial world ranging from viruses to pro and eukaryotes? We must be able to say with certainty that the detection of a certain organism means involvement in host damage in an appreciable number of instances; we must know the changes in our amphibiotic microbiota, our normal microbes that can become involved in disease production before we can attach decisive interpretations to our results. Of course, there are numerous exceptions that all of us appreciate. The microorganisms endowed with known roles in the etiology of specific diseases are the best candidates for the application of the emerging molecular and immunological technology, joined by thoses etiological agents that have escaped easy domestication to date.

Both molecular probes and monoclonal antibodies hold the promise of completely changing all aspects of the entire medical diagnostic laboratory service. Undoubtedly all divisions including pathology shall eventually employ similar technology, a technology that begs to be automated. Clinical microbiologists must recognize that these and even more sophisticated future technology present an unusual opportunity to assert their qualifications in the application of the scientific and clinical advantages provided by the emerging methods. Active participation in the evolution of these practices and repeated demonstration of the clinical microbiologists' traditional expertise to combine physical, chemical and biological principles for the benefit of the patient should insure the continued participation of clinical microbiologists in the realm of diagnosing disease (6).

The apprehensive response by laboratory scientists with respect to office, clinic and even bedside microbiological analyses must be considered

in light of the presently available technology and laboratory skills required
for the performance of even the simplest microbiological examinations (5,12).
As part of this consideration, we must appreciate that infectious disease in
the United States can be separated into two major categories; those acquired
in the community in contrast to infections of immunocompromised individuals,
mostly at risk in the institutional setting, albeit that patients in remis-
sion from their underlying disease and residing in the community may manifest
complications usually encountered in the hospital or involving microorganisms
rarely harmful to healthy community residents. Identification of the
bacteria, fungi, viruses and protozoa involved in complicating the diseases
of the immunocompromised individuals demands, at this time, the skill of
well-trained clinical microbiologists. Office and clinic identification of
select community etiological agents of infections is advocated increasingly.
Even a cursory examination of these "simple" tests reveals several interesting attributes. They are limited to upper respiratory infections, urinary
tract infections and the diagnosis of some sexually transmitted diseases.
Most of these tests, despite their touted simplicity, require several manip-
ulations and some incubation. Several examinations depend on equipment not
usually available in the office of the community physician. The available
examinations have been evaluated in clinical laboratories by staffs accus-
tomed to the performance of tests, the use of controls, the mind-set and
experience of the laboratorian and without the detractions present in the
office and the clinic.

Furthermore, the training of physicians and nurses during the recent
past has neglected the teaching of laboratory skills. Curricula in pro-
fessional schools have sacrificed bread and butter microbiology to the eso-
teric aspects of the DNA-RNA-protein sequence and the commendable emphasis
on the psychosocial welfare of the patient (3,4). Useful tools for physicians' offices and clinics must meet certain minimal requirements: reactions
must be completed immediately; new chromogenic indicators must be found that
separate positive and negative reactions unequivocally, a challenge that the
ingenuity of chemists undoubtedly can accomplish; the test must not present
an infection hazard to personnel during the performance of the test and when
discarded; the test must differentiate between the mere presence of an agent
and its critical concentration that suggests imminent overt disease. Considerable advances in microbiological technology are needed before appropriate and accurate tests can be available for office and clinic use.

Several approaches are excellent candidates for advancing microbiological analyses. Gas liquid and high performance liquid chromatography have
been applied successfully to the examination for microbial end products, con-
stitutuents and antimicrobic agents. Direct application of these examinations to specimens is hampered by the complexity of preparatory steps that
cannot be easily automated. However, both modalities should not be overlooked as appropriate detectors of peculiar gene products as our collective
understanding of species differences advances. Such comprehension of microbial activities may provide particular species specific enzymes that may
herald the presence of an organism directly in the specimen. Nuclear mag-
netic resonance analysis at this time lacks the sensitivity required to
differentiate particular compounds in specimens or culture supernatants.
Thus, molecular probes and monoclonal antibodies and possibly, at the risk
of heresy, a combination of both techniques hold the greatest promise to
attain our goal of making clinical microbiology a diagnostic discipline.
The introduction of these techniques should not be random, but proceed in an
orderly fashion. Since we do not readily identify a number of significant
bacteria, fungi, many protozoa and most viruses, tools for their recognition
should be the first offered. Gradually, more of the labor intensive tasks
of microbiological analysis could be replaced with innovative techniques that
eventually shall enable us to detect potentially harmful microorganisms in
specimens in addition to the enzymes or microbial constituents that provide

a profile of antibiotic susceptibility. I appreciate that this scenario represents a simplistic version. I am aware that certain polymicrobic specimens contain numerous amphibiotic representatives that provide an uninterpretable spectrum, no less confusing with these approaches than with our present methods. I hope that the recognition of the significant agent(s) can be aided by detecting them or typical products or constituents in blood or urine increasing the chance to properly sequester the most likely culprit. We should insist that the introduction of these new modalities be carried out against a background of clinical significance rather than insisting on comparison with the less accurate pure culture technique that does ignore the aforementioned anatomic attributes needed by the organism to succeed in a natural setting.

The survival of clinical microbiology demands our accommodation to the needs of the patient, of the need to be diagnostic (5). It means we must abandon the impediments of habit and tradition. We must structure an approach that weds our new and old knowledge of microbiology to answer the first and universal challenge: does the patient have an infection? Our efforts can and should be directed at that problem.

We have identified a variety of crystalloid molecules that are peculiar constituents and products of bacteria, not encountered in any other living organism. Similarly, particular compounds are encountered in yeasts, fungi and protozoa. Even viruses manifest their presence by directing host cells to synthesize their building blocks that differ from those found in the host fluids in their absence. Some of these peculiarly microbial materials can be identified in blood, urine or spinal fluid by gas liquid chromatography and high performance liquid chromatography. It should not be beyond our ingenuity to construct appropriate antibodies for the recognition of these materials made haptenic by combination with carrier materials. Such antibodies when applied to patient's bloods or urines could rapidly tell the presence or absence of an infection by one or more major microbial groups. Certainly, bacterial participation in an infectious process must exceed a certain threshold of such compounds, contributed by the normal microbiota. The benefits of knowing which of the major etiological agents' groups is involved in the patient's symptoms need not be belabored; the impact on initial therapy alone would be considerable. This information could be followed by the examination of an appropriate specimen. In most instances, we could proceed in a stepwise fashion that would lead us with the help of antibodies and molecular probes to ascertain with bacteria, for example, the gram characteristics, the major subdivisions and finally the identity of the etiological agent. Of course, we must learn to amplify our probes or to increase the copy numbers in the microorganisms in order to succeed with these advances. The same future technologies should aid in recognizing the potential antibiotic response of the organisms involved.

I appreciate that this approach is but one solution to the provision of rapid accurate diagnostic microbiology. It signifies only that we have tarried too long in waiting to advance our skills and capabilities. Instead of defending our ancient approaches, we should welcome and pursue any and all means that have the potential to improve our service to all of humanity.

LITERATURE CITED

1. Bartlett, R.C. 1982. Medical microbiology: how far to go - how fast to go in. 1982. In V. Lorian (ed) Significance of Medical Microbiology in the Care of Patients, 2nd ed. pp. 12-44. Williams and Wilkins, Baltimore.

2. Farmer, J.J., B.R. Davis, F.W. Hickman-Brenner, A. McWhorter, G.P. Huntley-Carter, M.A. Asbury, C. Riddle, H.G. Wathen-Grady, C. Elias, G.R. Fanning, A.G. Steigerwalt, C.M. O'Hara, G.K. Morris, P.B. Smith and D.J. Brenner. 1985. Biochemical identification of new species and biogroups of <u>Enterobacteriaceae</u> isolated from clinical specimens. J. Clin. Microbiol. 21:46-76.

3. Isenberg, H.D. 1979. <u>Legionella</u>, WIGA, et cetera: Pathogens? Ann. Int. Med. 91:785-786.

4. Isenberg, H.D. 1982. Microbiology and the ailing patient. In V. Lorian (ed) <u>Significance of Medical Microbiology in the Care of Patients</u>, 2nd ed. pp. 1-11. Williams and Wilkins, Baltimore.

5. Isenberg, H.D. 1987. Home, bedside and office microbiological tests: an enigmatic dilemma. In P.J. Howanitz (ed) <u>Quality Assurance in Physician Office, Bedside and Home Testing</u>. pp. 164-177. College of American Pathologists, Skokie.

6. Isenberg, H.D. 1988. Pathogenicity and virulence: another view. Clin. Microbiol. Rev. 1:40-53.

7. Isenberg, H.D. and J.I. Berkman. 1962. Microbial diagnosis in a general hospital. Ann. N.Y. Acad. Sci. 98-647-669.

8. Jones, R.N. 1982. The antimicrobial susceptibility test: rapid, overnight, agar and broth, automated and conventional, interpretation and trend analysis. In V. Lorian (ed) <u>Significance of Medical Microbiology in the Care of Patients</u>, 2nd ed. pp. 341-369. Williams and Wilkins, Baltimore.

9. Lennette, E.H., A. Balows, W.J. Hausler and H.J. Shadomy. 1985. <u>Manual of Clinical Microbiology</u>, 4th ed. American Society for Microbiology, Washington, D.C.

10. Louria, D.B. 1982. The speciation polemic: an analysis of the debate largely from the physician's point of view. In V. Lorian (ed) <u>Significance of Medical Microbiology in the Care of Patients</u>, 2nd ed. pp. 45-52. Williams and Wilkins, Baltimore.

11. McCabe, W.R. and K.D. Sottmeier. 1982. Clinical significance of speciation of aerobic gram negative bacilli. In V. Lorian (ed) <u>Significance of Medical Microbiology in the Care of Patients</u>, 2nd ed. pp. 53-63. Williams and Wilkins, Baltimore.

12. Needham, C.A. 1987. Rapid detection methods in microbiology: are they right for your office? Med. Clin. North Amer. 71:591-605.

13. Smith, T. 1934. <u>Parasitism and Disease</u>. Princeton University Press, Princeton.

14. Steel, K.J. 1962. The practice of bacterial identification. Symp. Soc. Gen. Microbiol. 12:405-432.

15. Wilson, G., A. Miles and M.T. Parker. 1983. <u>Topley and Wilson's Principles of Bacteriology, Virology and Immunology</u>, 7th ed. Williams Wilkins, Baltimore.

THE USE OF DNA PROBES TO DETECT AND IDENTIFY MICROORGANISMS

David E. Kohne

Chief Scientist
Gen-Probe Incorporated
San Diego, California

INTRODUCTION

The objective of this article is to provide the reader a basis for understanding what DNA probes are and how they can be used to detect microorganisms.

The general theme of this presentation is largely a result of my basic research interest over the last 15 years; which has been to describe the bacterial and viral flora of normal humans in order to assess the disease potential which humans carry with them. At the beginning of this study I searched for the best technical approach and decided to detect the microorganisms by detecting their nucleic acids using the process of nucleic acid hybridization. The advantages of this approach include the following points:

1. If the target organism is present in the sample, the nucleic acids of the target organism must be present, no matter what the state of genetic expression of the organism in that sample.

2. A nucleic acid sequence specific only for a particular microorganism can always be found.

3. Nucleic acid probes highly specific for narrow or broad groups of target organisms can be developed.

4. The nucleic acid hybridization approach is highly sensitive and has the potential of detecting a single organism.

At that time, however, DNA probes had been little used to detect organisms in clinical samples. In addition, sample preparation which required nucleic acid isolation and purification, and the hybridization assay were both lengthy, labor-intense processes. Many of the DNA probe methods used today still have these disadvantages.

Makeup of Nucleic Acids

DNA or RNA probes provide a powerful tool for the detection, identification and quantitation of microorganisms. The key property of DNA probes is that specific DNA probe molecules can recognize and selectively interact with target nucleic acid sequences present only in the target organism. This process of recognition and interaction is called nucleic acid hybridization.

Let's briefly examine the molecular basis which makes it possible to use specific nucleic acid molecules as DNA probes.

Nucleic acids are polymers composed of four basic subunits (also called bases) which are covalently linked in a linear fashion. Two types of nucleic acid are present in nature. DNA, which is the genetic material or the genes, and RNA which is a direct gene product.

DNA in nature is almost always double stranded and is composed of 4 subunits or bases. These are Adenine, Guanine, Cytosine and Thymine; A, G, C, and T. The individual strands of the double-strand molecule are held together by specific interactions between the bases in each strand. A and T specifically interact together as do G and C. AT and GC are called complementary base pairs. In the double-strand molecule, each base in one strand is paired with its complementary base in the other strand. Each A is paired with T and each G is paired with C. The chemical glue which holds the double strands together is made up of the many complementary base pair interactions. The individual single-strand molecules which make up the double-strand molecule are called complementary single strands.

RNA is also composed of 4 subunits or bases, Adenine, Guanine, Cytosine and Uracil (rA, rG, rC, rU). Complementary RNA molecules can recognize each other and interact to form a double strand form which is held together by interactions between the complementary bases rArU and rGrC. RNA in nature is almost always single strand in form. Only one of the two complementary strands is present in the cell. RNA is almost always a direct gene product with DNA serving as a template for the enzymatic synthesis of RNA.

Complementarity can also exist between RNA and DNA molecules resulting in a double-strand molecule which is composed of one strand of RNA and one strand of DNA. Again, the molecules are held together by interactions between complementary base pairs. In this case, the pairs can be rAT, ArU, rGC, GrC.

For perspective, the number of bases present in the DNA of different organisms is presented in Table 1.

TABLE 1

NUMBER OF BASES IN THE DNA SEQUENCES
OF AN ORGANISM VARIES

Mammal	7×10^9	bases, 23 chromosomes
Bacteria	10^7	bases, 1 chromosome
Large virus (HSV)	3×10^5	bases, 1 chromosome
Small virus (Polio)	8×10^3	bases, 1 chromosome
Smallest virus (viroid)	300	bases, 1 chromosome

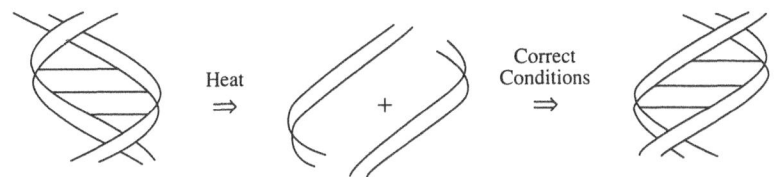

- *E. coli* DNA contains 10^7 bases
- Each strand contains 5×10^6 bases
- Any complementary strand can hybridize to any other complementary strand

Figure 1. Schematic Representation of Hybridization

NUCLEIC ACID HYBRIDIZATION

The principle behind DNA probe technology is nucleic acid hybridization (1,2): Hybridization is the process of formation of stable
double-strand nucleic acid molecules from complementary single-strand
molecules. Nucleic acid hybridization is a rational chemical reaction and
can be used in a predictable manner to detect and quantitate microorganisms.

Figure 1 presents a schematic illustration of hybridization using
E. coli DNA as an example. E. coli DNA contains 10^7 bases. Since it is a
double-strand molecule, each single strand is composed of 5 x 10^6 bases,
and each base in one strand is paired with its complementary base in the
other strand.

The double-strand nucleic acid molecule can be converted to two
single-strand molecules by a variety of methods, heat being one. These
complementary single strands wander around free in solution, and
eventually a single-strand molecule will collide with its complementary
molecule, chemical recognition (i.e., sequence recognition) will occur,
and the two strands will interact to form a stable double-strand
molecule. This process of collision, chemical recognition and interaction
between two complementary strands is the process of nucleic acid hybridization. The basis for the selectivity of DNA probes lies in the ability
of complementary sequences to specifically recognize each other and form a
stable double-strand molecule.

IDEAL AND NON-IDEAL NUCLEIC ACID HYBRIDIZATION

Must two nucleic acid molecules be perfectly complementary in order
to hybridize together? The answer is no. There are two different types
of non-ideal hybridization. However, it will be useful to first describe
ideal hybridization, again using E. coli DNA as an example.

E. coli DNA contains 10^7 bases and is a double-strand molecule. Each
single-strand DNA molecule contains 5 x 10^6 bases. Each of the 5 x 10^6
bases in one strand is properly matched with its complementary base in the
other strand resulting in: a) perfect end-to-end matching of the complementary DNA strands, and b) perfect base pair matching between the bases
in each strand.

This double strand form can be converted to two complementary single-
strand molecules by raising the temperature of the solution to a point
where the double strand form is unstable. Under the proper conditions,

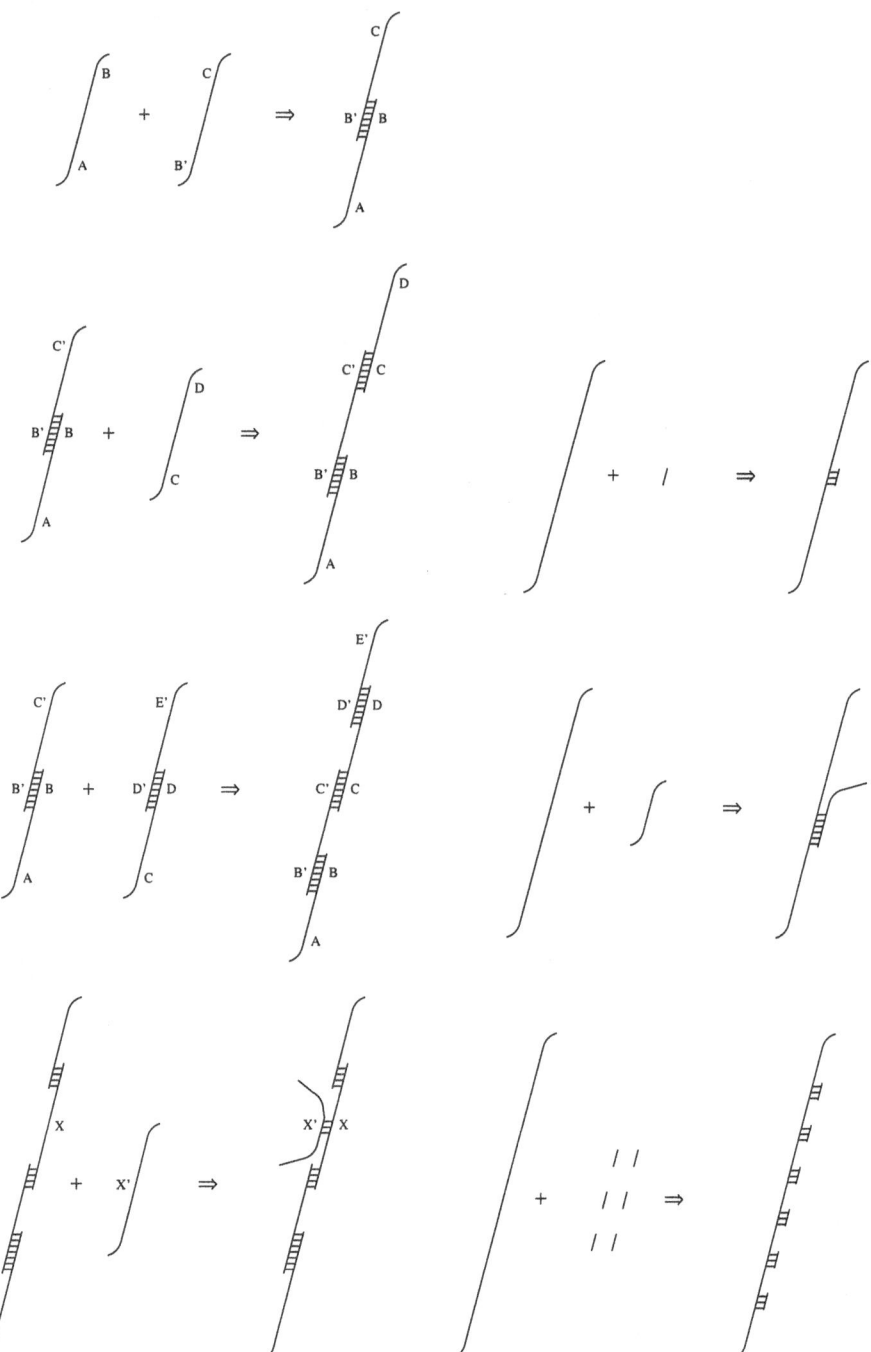

Figure 2a. Imperfect End to End
Hybridization

Figure 2b.

these complementary single-strand molecules can chemically recognize each other and interact to form a stable double-strand molecule where the individual single strands are perfectly matched end to end, and all of the bases in one strand are paired with their proper complementary base in the other strand. This is ideal hybridization.

One type of non-ideal hybridization is imperfect end-to-end matching as illustrated in Figure 2. Two nucleic acid molecules which have complementary and non-complementary regions can hybridize together to form a partially double, partially single-strand hybrid molecule. A large number of permutations can occur under this format and many of these are illustrated in Figures 2A and 2B. Figure 2A illustrates various types of hybrids often formed when double-strand DNA is fragmented and hybridized. Figure 2B illustrates hybrids which are often formed by hybridizing small DNA probes to large target molecules. For example, a 30-base long DNA probe can hybridize with a 3000-base long RNA target molecule resulting in a hybrid in which the DNA probe is hybridized 100 percent, but only 1 percent of the RNA is in a hybrid state. At the other extreme, a probe mix consisting of one hundred different 30-base long DNA probes, each complementary to a different region of the 3000-base long molecule, can hybridize to the large molecule resulting in the entire target molecule being hybridized.

A second type of non-ideal hybridization results in imperfect base pair matching in the double strand region of a hybrid. This is illustrated in Figure 3. Two nucleic acid molecules can have perfect end-to-end matching but not have perfect sequence complementarity. Under the proper conditions, these partially complementary molecules can hybridize together to form a stable double-strand hybrid which is perfectly matched end to end but contains base pair mismatches. The imperfect base pair matching results in a double-strand molecule which is less stable (i.e., converts to a single strand form at a lower temperature) than if perfect base pair

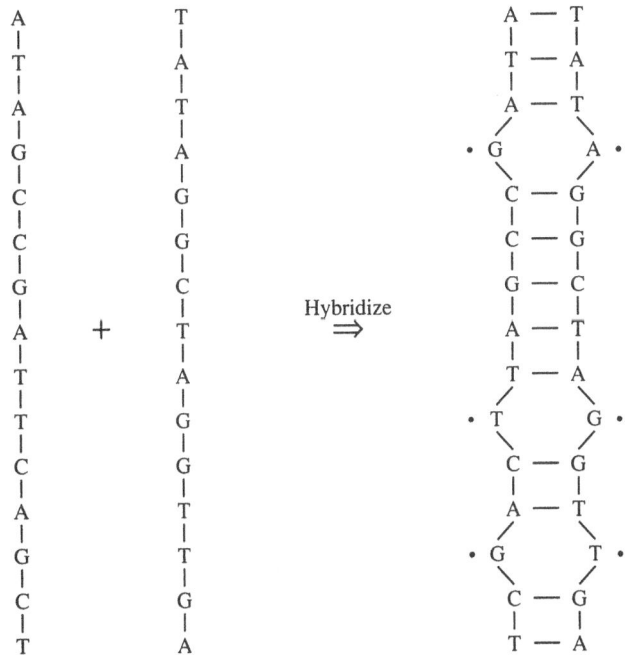

Figure 3. Imperfect Base Pair Matching

matching were present since the molecule with base pair mismatches has less "chemical glue" holding the two strands together. It is possible to get highly specific hybridization between partially complementary single strands resulting in double-strand molecules with 20-30% base pair mismatches in the double-strand region. Furthermore, it is possible for a hybridized molecule to have both imperfect end-to-end matching and imperfect base pair matching. This ability of partially complementary single-strand molecules to hybridize greatly enhances the power of DNA probes as a diagnostic tool. This will be illustrated below.

When put under the proper conditions, two nucleic acid molecules which are only partially complementary can hybridize together to form a stable double-strand molecule. What are these conditions? In order for complementary single-strand molecules to hybridize together, the resulting double-strand molecule must be stable under the conditions of hybridization. Therefore, the hybridization must be done at a temperature below the temperature at which the double-strand form is converted to single-strands.

How much below the dissociation temperature do you want to go? Hybridized molecules which have no mispaired bases have the highest dissociation temperature. If the hybridization is done at a temperature just below this dissociation temperature, then only hybrids with a high degree of base pair matching can form. However, if hybrids with a greater degree of mispaired bases are desired, then the temperature of hybridization is lowered. The lower the hybridization temperature, the greater the degree of base pair mismatches which can occur in the resulting hybrids. Thus by changing the hybridization temperature, one can select the degree of base pair matching which is required to form a stable hybrid.

This illustrates a very important property of DNA probes: the specificity of many DNA probes can be changed by simply changing the conditions of hybridization. The same DNA probe sequence can have broad or narrow specificity depending on the conditions of hybridization. When hybridized at higher temperature, it has narrow specificity and will detect only a limited number of related organism types. When hybridized at lower temperature, it has broader specificity and will detect a larger number of related organism types. In practical terms, this means that the same probe may be used at low temperature for screening purposes and at high temperature for a more precise identification.

THERMAL STABILITY OF HYBRIDIZED PROBE

The thermal stability (TS) of a Probe:Target hybrid is a measure of the temperature at which the double-strand hybrid is converted to single-strand molecules. TS is defined as the temperature at which 50 percent of the hybridized probe is converted to single-strand form. This can be measured by taking preformed hybrids and heating them to increasingly higher temperatures. After heating to each temperature, an assay is done to determine the fraction of probe which has been converted to the single-strand form.

Earlier, it was stated that DNA probes can have either broad or narrow specificity depending on the conditions of use. Whether a probe is useful to detect only a specific type of organism depends largely on the TS difference between Probe:Target (PT) hybrids and any Probe:Non-Target (PNT) hybrids which might be formed. What is wanted is a temperature of incubation where stable PT hybrids will form while no PNT hybrids will form because they are unstable. The (TS (PT) - TS (PNT)) difference should be as large as possible. The larger the difference, the easier it is to design the test. This is the basic principle behind the design of

Normal sequence
in parent

. . . . . A G C A G C T A G C T A G C T A G C T A G C . . . . .

           ↑

Mutational Event

Mutant sequence
in offspring

. . . . . A G C A G C T A G C T C G C T A G C T A G C . . . . .

           ↑

This change can result in a
disease such as:
     β. Thalesemia
     Sickle Cell Anemia
     Certain Cancers

Figure 4a. Many Diseases Are Caused by a Single Base Change in the
Sequence of the DNA of an Otherwise Normal Individual

all DNA probe diagnostic tests. Consequently, factors which affect the TS
can affect the probe specificity, and these factors must be controlled
when doing a DNA probe test.

To illustrate this principle, let us examine a DNA probe test
designed to provide genetic information which can be used for human
genetic counseling (3). Many diseases are caused by a single base change
in the sequence of an otherwise normal individual. This can result in a
genetic disease such as beta thalesemia or sickle cell anemia, and the
onset of certain cancers can be triggered by such a single base mutation.

Figure 4A illustrates such a change. DNA probes can be used to
detect a single specific base change which has occurred in the DNA. To
accomplish this one first must know the sequence of the normal and mutant
gene sequences. Then a short probe can be produced which is 100 percent
complementary to the mutant sequence (Figure 4B) but has a single base
pair mismatch with the normal or wild type DNA sequence (Figure 4C). The
TS of the perfectly matched hybrid (40°C) is 10°C higher than that of the
single mismatch hybrid (30°C) (Figure 4D). When hybridized with fetus DNA
at 35°C only perfectly matched hybrids are stable and can form. Thus if
this probe hybridizes at 35°C with fetus DNA then the fetus carries the
mutant gene sequence.

Probe sequence will form a
perfect match with the mutant

Target
Sequence

. . . . . A G C A G C T A G C T C G C T A G C T A G C . . . . .

        C G A T C G A G C G A T C G A T C G

                  ↑

Probe
Sequence   Perfect Match

Figure 4b. DNA Probes Can Be Used to Detect a Specific Base
Change Which Has Occurred in the DNA

Normal
Sequence

. . . . . A G C A G C T A G C T A G C T A G C T A G C . . . . .
C G A T C G A G C G A T C G A T C G
Probe   ↑
Sequence   Mismatch

Figure 4c. Short Probe Sequence Will Form an Imperfect Match
with the Normal Sequence

Hybrid	Number of Mispaired Bases	Thermal Stability
Probe:Mutant	0	40° C
Probe:Normal	1	30° C

1. When hybridized at 35° C only perfectly matched hybrids can form.

2. If this probe hybridizes at 35° C with DNA from a fetus, then the fetus carries the mutant gene.

Figure 4d. Thermal Stability of Probe and Probe Mutant Hybrids

SIMILARITY OF NUCLEIC ACID SEQUENCES AMONG DIFFERENT ORGANISM TYPES

The more closely related two organisms are, the more similar are their nucleic acid sequences (4,5). DNA sequences in members of the same species are virtually identical while DNA's from distantly related organisms have very little sequence similarity. The DNA's of man and chimp cross-hybridize to greater than 90% with 1-2% base pair mismatches while DNA's from man and mouse cross hybridize less than 5%. Among bacteria, the DNA's of Salmonella typh. and E. coli cross hybridize about 60% with a moderate number of base pair mismatches while DNA's from E. coli and Legionella bacteria cross hybridize very little.

Nucleic acid sequence specificity provides a powerful and general means for detecting and identifying organisms of a given type since: all organisms contain nucleic acids; a nucleic acid sequence specific for a given organism type can always be found; a specific probe can be developed for any specific sequence.

COMPONENTS OF DNA PROBE TEST

There are four basic components to a DNA probe test.

PROBE   -   a labeled single-strand sequence complementary
                      to the target sequence to be detected.

SAMPLE   -   ranges from water to feces or sputum and may or
                      may not contain the target organism.

HYBRIDIZATION   -   the conditions of solution composition and
METHOD   temperature under which hybridization occurs.

HYBRIDIZATION   -   a method for detecting the presence of Probe:
DETECTION   Target hybrids
METHOD

The first two components will be discussed in some detail below.

GENERAL CHARACTERISTICS OF PROBES

An ideal probe is single-strand DNA or RNA which can hybridize only if the target is present in the sample. Such a probe contains only one of the complementary strands necessary for hybridization and when incubated alone, cannot form double-strand molecules because only one of the complements is present. The size of a probe can range 10 to over 10^4 bases in length. The length chosen depends on the specific requirements of the test.

Probe specificity lies in the probe sequence and the conditions under which it is used, as discussed earlier. The specificity requirements are that the probe must hybridize with the target nucleic acid but not hybridize with any non-target nucleic acids which may be present in a sample. Non-target nucleic acids which may be present in a human clinical sample are human and any contaminating microorganism nucleic acids. When properly designed and used a DNA probe has essentially absolute specificity and will hybridize only to target nucleic acid. In order to be useful as a diagnostic a probe need not have absolute specificity for the target organism. For example, the DNA probe used in Gen-Probe's diagnostic kit for Mycoplasma pneumonia will hybridize to Mycoplasma genitalium as well as the target. This does not affect the use of the probe detecting M. pneumonia in throat swab samples, however, since M. genitalium has never been found in the respiratory tract.

Probe must be marked either directly or indirectly in order to follow its fate. DNA probes can be marked with either radioactive (^{32}P, ^{35}S, ^{3}H, ^{14}C, ^{125}I) or non-radioactive (biotin-avidin systems, hapten-antibody systems, fluorescent, chemiluminescent) reporter molecules. In general, anything which can be used to mark antibodies can be used to mark probes.

Probes can be selected by determining the base sequence of target and non-target DNA or RNA, then comparing the sequences and selecting an appropriate sequence which is characteristic only of the target nucleic acids. Once the probe sequence is selected, the probe can be produced with a DNA synthesizer. Probes can also be selected on the basis of the hybridization characteristics desired. In this case, neither the probe sequence nor its function need be known.

The properly designed and used DNA probe can detect any type of genetic change. These include: single base changes as discussed earlier; multiple base changes; rearrangements of all kinds; additions and deletions; changes in genetic expression; and quantitative changes in genetic expression.

It should be emphasized that both the probe and the target sequences must be single stranded at the start of a DNA probe test in order for the test to work.

A PROBE TO POLIO I RNA: BASICS OF PROBE SELECTION

This example is derived from a research project the author was involved in some time ago. The object of the project was to determine whether normal and diseased human tissue contained RNA sequences similar to the sequence of Polio virus I RNA (6). A radioactive probe complementary to Polio RNA was produced by using Polio I RNA as a template for the enzyme reverse transcriptase to synthesize small pieces of radioactive DNA complementary (cDNA) to Polio I RNA. Each small piece of cDNA represented a small fraction of the Polio I RNA sequence, but the population of cDNA pieces represented the entire Polio RNA sequence.

TABLE 2

CHARACTERIZATION OF ^{3}H cDNA

10^{-5} microgram	Probe + Poly A	0.1%
10^{-5} microgram	Probe + 0	0.1%
10^{-5} microgram	Probe + 200 Micrograms Human RNA	0.1%
10^{-5} microgram	Probe + Mix of 200 Micrograms Human RNA and 0.1 Microgram Polio I RNA	99%

The hybridization characterization of this cDNA probe is shown in Table 2. The probe does not self hybridize and when incubated alone only 0.1% of the probe behaves as if hybridized. This 0.1% is not true hybridization but represents the non-specific background signal due to the assay. The negative control shows that probe incubated with human RNA also does not hybridize, even at a ratio of 10^7 parts human RNA to 1 part probe. The same value would be obtained if the amount of RNA were 20 milligrams or 2 micrograms, and if RNA's or DNA's from widely different bacterial or mammalian sources were present in the hybridization mix. This gives an indication of the extremely high specificity of probes. The probe collides frequently with the heterologous nucleic acid but does not chemically recognize and specifically interact with them to form a stable double-strand hybrid. The positive control illustrates this also. The probe hybridizes almost completely with Polio I RNA even in the presence of a large excess of heterologous nucleic acid. The presence of the heterologous nucleic acid changes neither the kinetics nor extent of probe hybridization.

TABLE 3

SPECIFICITY OF POLIO I 3H cDNA
PROBE HYBRIDIZATION

		HYBRIDIZATION EXTENT OF PROBE
Polio I Probe	+ Polio I RNA	99%
Polio I Probe	+ Polio I RNA (5 Strains)	95-99%
Polio I Probe	+ Polio II RNA (8 Strains)	77-80%
Polio I Probe	+ Polio III RNA (8 Strains)	69-78%
Polio I Probe	+ Coxsackie RNA	5%
Polio I Probe	+ EMC RNA	3-5%
Polio I Probe	+ VSV RNA	0.1%
Polio I Probe	+ MS-2 RNA	0.1%

Probe for Polio I RNA	1'	2'	3'	4'	5'	6'	7'	8'	9'	10'
Polio I RNA	1	2	3	4	5	6	7	8	9	10
Polio II RNA	1	2	3	4	5	6	7	11	12	13
Polio III RNA	1	2	3	4	5	6	14	15	16	17
Coxsackie RNA	1	18	19	20	21	22	23	24	25	26
EMC RNA	1	27	28	29	30	31	32	33	34	35

Figure 5. Idealized Pattern of Sequence Relationships Among Related Viruses

This probe then is ready to use for viral detection. It does not hybridize with human cell nucleic acid but does hybridize to Polio I RNA. At this point, we must ask whether this probe hybridizes to any other virus nucleic acid besides Polio I RNA? The answer is yes it does, as can be seen in Table 3. This probe hybridizes to varying extent with all member viruses of the related group of viruses called "Enteroviruses" but does not hybridize to viruses which are not members of this group. Thus this cDNA probe, which represents all of the Polio I RNA sequence, can be used to detect a large number of different but related viruses. Such a probe can be used for screening samples for the presence of any member of the Enterovirus group. However, only in special circumstances can this cDNA probe be used to identify the type of Enterovirus present in a sample.

This raises another question. Can a probe which detects only Polio I RNA be produced?

The answer is yes. To illustrate this, it is necessary to examine an idealized pattern of nucleic acid sequence relationships among related viruses (Figure 5). It should be kept in mind that this pattern of sequence relationships is the same general pattern seen for nucleic acid sequence relationships among all related groups of organisms.

In this case, each virus RNA is segmented into 10 different sub-sequences. Each Enterovirus RNA is about 8000 bases long. A DNA probe for Polio I RNA contains 10 different subsequences, each of which is complementary to a different subsequence of the Polio RNA. By definition, this probe will hybridize to 100 percent with Polio I RNA. However, this probe cannot hybridize completely with any other Enterovirus RNA since the probe contains DNA subsequences (8, 9 and 10) which can only hybridize with Polio I RNA.

Probe complementary only to Polio I subsequences 8 or 9 or 10, or a mixture of 8, 9 and 10, will detect only Polio I RNA. This illustrates the basic pattern of nucleic acid sequence similarity seen for all related groups of organisms. Some sequences are held in common. Other sequences are not held in common, and these are the sequences which are character-istic of a specific group. Probes made to these sequences will detect only that group.

A probe specific only for each of the other Enteroviruses can also be produced by following this same principle. A variety of methods are available for selecting the proper subsequence to target.

If a probe were desired which would hybridize completely with all Enteroviruses, then subsequence 1 would be the chosen target. This would be a useful screening probe.

21

The strategy for designing a DNA probe assay is very important and basically involves choosing the appropriate target and probe molecules with which to work. The conventional strategy involves using a DNA or RNA probe to detect a target DNA or gene sequence. Gen-Probe's strategy is to use DNA and RNA probes to target RNA sequences of the cell as a general class. Our initial efforts have focused primarily on using ribosomal RNA (rRNA) as the target molecule.

Ribosomes are present in all cellular life forms. There are _always_ ribosomes present in a cell as they are an integral part of the protein synthetic machinery. Both procaryotic and eucaryotic cells contain large numbers of ribosomes and, therefore, rRNA molecules. rRNA is a direct gene product, and one strand of the rRNA gene is used as a template to produce rRNA which is then incorporated into a ribosome. A ribosome is composed of about 50% protein and 50% rRNA. In a typical bacterial cell such as E. coli there are about 10^4 ribosomes. The rRNA is in a single-strand form and RNA complementary to the rRNA is not present in the cell.

There are several dramatic advantage to this approach. One is a greatly enhanced (several thousandfold) sensitivity of detection. The other is that it allows the production of probes with specificities which were heretofore not attainable. These points are discussed in Tables 4 and 5.

There is a basic rationale for producing these probes targeted for rRNA, and it is based on the evolutionary history of the rRNA sequence. The total primary base sequence of rRNA in bacteria is about 4600 bases long. This sequence is generally considered to have been highly conserved during evolution (5). The rRNA sequence can be viewed as being composed of a population of short subsequences. During evolution, certain of these subsequences have been highly conserved and are present in all cellular

TABLE 4

THE ENHANCED SENSITIVITY OF THE
GEN-PROBE RIBOSOMAL RNA
DETECTION METHOD

==

    o   rRNA and DNA can be detected equally well by Nucleic Acid hybridization.

    o   An E. coli cell contains:

        10^4 ribosomes
        10^4 rRNA's
        2.5×10^{-14} grams rRNA

    o   That same E. coli cell contains:

        4 genes which code for rRNA
        10^{-17} grams of ribosomal DNA

Thus, by choosing rRNA as the target rather than DNA, the detection sensitivity is increased several thousandfold.

==

22

TABLE 5

SPECIFICITY OF DNA PROBES TO rRNA

1. The method allows probes with widely varying specifi-
    cities to be produced. This allows:

    A. The detection of the presence of any life form
        (excluding viruses) in a sample, with the
        performance of a single lab assay.

    B. The specific detection of any member of a broadly
        related class (e.g., all bacteria) of organisms
        in a sample with just one lab assay, even in the
        presence of large numbers of unrelated organisms.

    C. A battery of different DNA probes to be used for
        the more precise identification (even down to the
        species level) of the organism present in the
        sample, even in the presence of large numbers of
        other organisms.

    D. The ability to quantitate the number of organisms of
        a specific type even in the presence of other
        organisms.

organisms. Probes targeted to these sequences will detect any cellular
life form. The remaining subsequences have diverged during evolution, and
some of these subsequences have diverged faster than others.

The key to producing probes with the unique specificities described
is to select the proper rRNA subsequences to serve as a target. Gen-Probe
has developed a variety of methods for selecting only those rRNA subsequences which are unique to the group of bacteria of interest.

Figure 6 presents the structure of the small (16s) ribosomal subunit
of E. coli. The heavy lines denote the sequences which have been highly
conserved during evolution. These are interspersed among the changeable
regions.

SAMPLE HANDLING

The basic requirement for sample handling is that the sample nucleic
acid must be made available for hybridization with the probe. The conventional approach for sample handling involves immobilizing the target on an
inert support. This requires at least a partial purification of the
sample nucleic acid.

The Gen-Probe approach to sample handling is quite different and
involves no immobilization step and no nucleic acid purification. The
sample is mixed directly with a probe solution which contains the probe;
lysing agents and solubilizer; nuclease inhibitors; and hybridization
promoters. In certain cases, such as Mycobacteria, some sample handling
steps may be necessary before mixing the probe solution and sample.
Sample types which we have assayed using this approach are: lung and
tissue homogenates; tissue culture medias; suspensions containing
bacterial or mammalian cells; bacterial pellets; urine; feces; sputum;
serum.

23

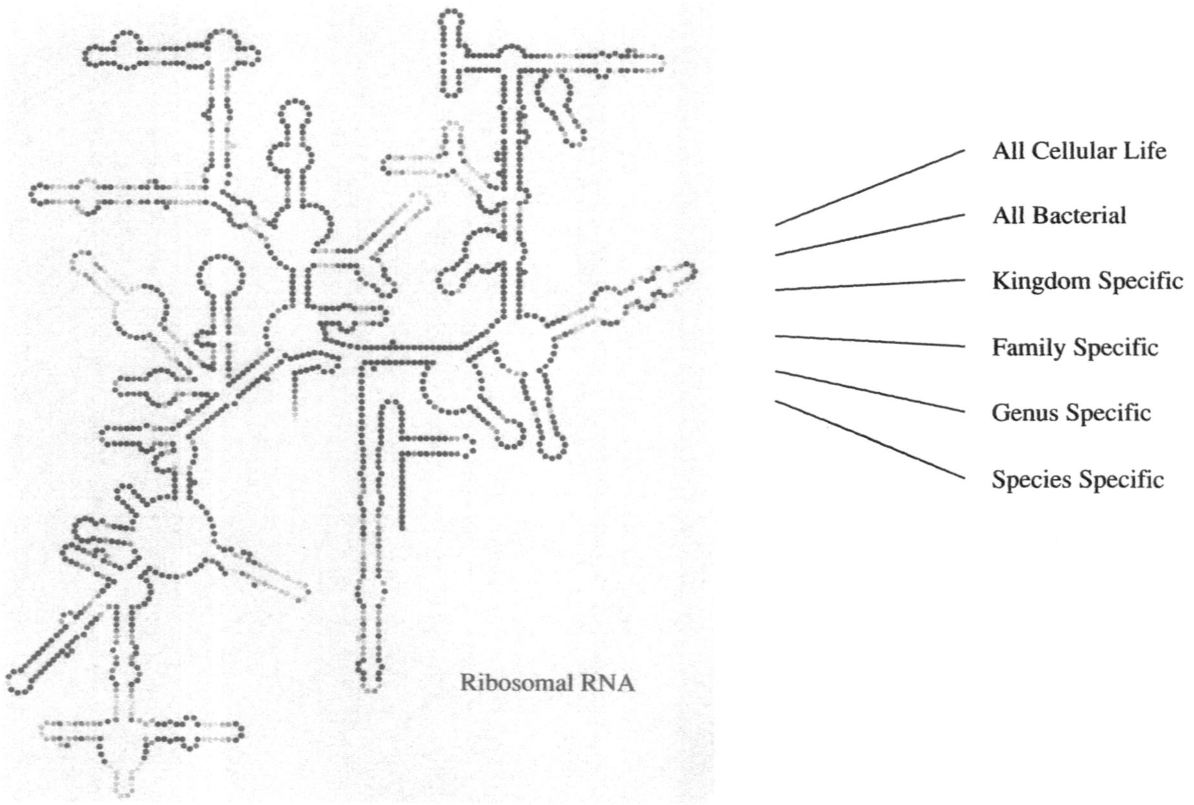

All Cellular Life

All Bacterial

Kingdom Specific

Family Specific

Genus Specific

Species Specific

Ribosomal RNA

Figure 6. Specificity

TABLE 6

RNA:DNA HYBRIDIZATION IN HIGH
CONCENTRATION OF CLINICAL SAMPLE

Clinical Sample	Concentration in Hybridization Mixture (Volume Percent)	% Hybridization of Probe
Sputum	56%	>90%
Feces	27% Solids	>90%
Serum (calf)	50% Serum	>90%
Legionella Infected Lung Homogenate	9% Lung Tissue	>90%

TABLE 7

RNA:DNA HYBRIDIZATION IN FECES

Sample	Probe	Percent Probe Hybridized
No feces, only bacteriophage MS-2 RNA	+ DNA complementary to bacteriophage MS-2 RNA	83%
Feces (final concentration of 27% solids) plus bacteriophage MS-2 RNA	+ DNA complementary to bacteriophage MS-2 RNA	83%
Feces (final concentration of 30% solids)	+ DNA complementary to bacteriophage MS-2 RNA	0.3%
No feces and no MS-2 RNA	+ DNA complementary to bacteriophage MS-2 RNA	0.3%
Feces (final concentration of 13% solids)	+ DNA complementary to E. coli rRNA	59%

25

DNA probes can be used in the presence of high concentrations of clinical samples. Table 6 summarizes data from such experiments. Table 7 presents details of RNA:DNA hybridization in high concentrations of feces. A probe which was partially damaged was used in these experiments in order to amplify any effect of the feces on the hybridization reaction. The same extent of hybridization was seen with or without feces present.

In order to rule out the possibility that the probe has bound to some feces component in a non-sequence dependent manner, probe was mixed with feces and no target RNA was added. None of the probe hybridized in this case. This demonstrated that the 83 percent signal was due to sequence-dependent hybridization of the probe to target RNA. The last result in Table 7 shows that rRNA endogenous to the feces can hybridize to a probe complementary to rRNA.

The sample handling step is the most difficult part of developing a DNA probe test. A variety of problems complicate this step.

Some samples, such as sputum, are highly viscous, and these samples must be liquified for easy handling. Hybridization inhibitors are present in many samples and can reduce the sensitivity of the test. Ribonuclease is in all samples and if the target is RNA, one must be able to keep it intact. Many organisms such as Mycobacteria, certain gram positives and yeast are difficult to break open to release the rRNA. Non-specific

TABLE 8

DNA PROBE TESTS FROM GEN-PROBE

	TEST	PROBE SPECIFICTY	MARKER
1.	Legionella Genus Culture Confirmation and Direct Specimen test	Genus	125_I
2.	Mycoplasma Direct	Species	125_I
3.	Mycobacterium Avium/Intracellullare Culture Confirmation	Species	125_I
4.	Mycobacterium tuberculosis Complex Culture Confirmation and Direct	Complex	125_I
5.	Mycobacteria Genus Culture Confirmation and Direct	Genus	125_I
6.	Chlamydia trachomatis Direct	Species	125_I, Chemi-luminescent
7.	Neisseria gonorrhoeae Culture Confirmation and Direct	Species	Chemilumi-nescent

binding (non-sequence dependent binding) of probe to a sample component can give false positive results.

Gen-Probe has been able to solve all of these problems and has incorporated the solutions into the test kits which are available.

DNA PROBE TESTS AVAILABLE

Gen-Probe has used its innovative DNA probe technology to produce simple, rapid and highly specific DNA probe tests for medically signifi-cant microorganisms. The first DNA probe test ever cleared by the FDA was a Gen-Probe test. Since that time, the FDA has cleared 15 DNA probe tests with 11 being Gen-Probe tests. Table 8 lists some of these tests. Certain tests employ isotopic labels while others utilize non-isotopic chemiluminescent markers. Culture confirmation tests are performed by assaying a suspension of cells made by putting a loop of a bacterial colony into water or saline. The direct tests are done directly on the clinical sample.

BASIC TEST FORMAT

The basic steps involved in these tests are: sample preparation, cell breakage, hybridization, separation of non-hybridized probe from hybridized, detection of hybridized marker group. Following is a skele-ton description of a DNA probe test for Mycobacteria.

1. Sample Preparation

    a) Add 100 microliters of sample solution to tube containing glass beads.

2. Cell Breakage

    a) Place tube in ultrasonic cleaner for 15 minutes in order to break open bacteria and free rRNA.

3. Hybridization

    a) Add 1 ml probe solution to tube. Incubate 72°C for 1 hour.

    b) Hybridization times can vary from 15 minutes to 4 hours depending on the test.

4. Separation

    a) Add 4 ml of separation mix. Incubate at 72°C for 5 minutes.

    b) Centrifuge at 2000 RPM for 2 minutes. Decant supernatant.

    c) Add 4 ml wash solution, centrifuge at 2000 RPM for 2 minutes and decant supernatant.

    d) Separation mix contains a separation agent. Hybridized probe binds to the separation agent and non-hybridized probe does not and can be washed away. Separation agent in tests is either hydroxyapatite crystals which are centrifuged out of solution, or magnetic particles which are pulled out of solution by magnets.

TABLE 9

A LARGE NUMBER OF PEOPLE IN A
HOSPITAL CAME DOWN WITH TB

1. The outbreak was traced to a patient admitted to the
   hospital with a hip abscess.

2. Biopsy was taken from abscess on hip and fixed in 10%
   formalin, processed with alcohols and xylene and
   embedded in paraffin.

3. Special stain showed tissue swarming with AFB.

4. Initial culture results showed only an avium-intracellulare complex bacteria.

5. Looked further and also isolated a TB bacteria.

6. Vast majority of organisms in culture were A-I
   organisms. Very few TB.

QUESTION: WAS THIS A MIXED INFECTION?

5. Detection

   a) Place tube in gamma counter and count 1 minute. Alterna-
      tively place tube in luminometer and measure light output.

AN EPIDEMIOLOGICAL PROBLEM

DNA probes can be very useful in certain epidemiological situations.
Table 9 summarizes an epidemiological case where probes were used to
resolve some questions concerning the epidemiology of the disease outbreak.

Portions of the formalin-fixed paraffin-embedded tissue were sent to
the author. The samples were processed to remove the paraffin and formalin
and put through a hybridization protocol similar to that described above
for Mycobacteria. Aliquots of the processed tissue were hybridized with
various Mycobacteria DNA probes as well as several control probes. Table
10 presents the results of this analysis.

Only two of the probes hybridized. Both the Mycobacteria Genus probe
and the Tuberculosis complex probe hybridized well to the tissue preparation. Four other probes, including probes for M. avium and M. intracellulare, did not detectably hybridize. The lack of hybridization of
these probes with the tissue preparation rules out any possibility of
non-specific binding or non-sequence dependent binding of DNA probes to
the tissue. This indicates that the signal seen with both the Genus and
TB probes is true hybridization.

The process of nucleic acid hybridization can be used to quantitate
the number of bacteria present as well as detect their presence. An
estimate of the number of Mycobacteria per gram of tissue was obtained by
hybridization kinetic analysis. The results indicated that the tissue
contained about $3-4 \times 10^9$ TB bacteria per gram of tissue. The same type

TABLE 10

RESULTS OF TISSUE ANALYSIS

	HYBRIDIZATION WITH TISSUE	% HYBRIDIZATION OF PROBE WITH TISSUE
Mycobacteria Genus Specific Probe	+	66%
M. avium Specific Probe	-	<1%
M. intracellulare Specific Probe	-	<1%
Legionella Genus Specific Probe	-	<1%
Vesicular Stomatitis Virus Specific Probe	-	<1%
TB Complex Specific Probe	+	80%

Conclusion:

1. The tissue nucleic acids hybridized only with Mycobacteria genus specific probe and TB complex specific probe.

2. No detectable hybridization of M. avium or M. intracellulare probes was obtained.

of analysis indicated that there was at least 1000 times more TB bacteria in the tissue than M. avium-intracellulare complex bacteria.

From this data, it is clear that the hip abscess contained a preponderance of TB bacteria which then caused the TB epidemic.

This study was done in collaboration with Mary Hutton, George Kubica, Bob Good, Vella Silcox, Charles Woodley, Margaret Floyd, Alan Block, Bill Jones of the CDC and William Stead of the Arkansas State Public Health Service.

TROUBLESHOOTING

DNA probes, when properly designed and used, have essentially absolute specificity. Convincing microbiologists that this is true is a challenge for Gen-Probe as a young company. This process of education involves comparing results from probe tests to culture, the current gold standard for most labs. Our customer service department occasionally receives reports from customers regarding samples which test positive with the probe and are culture negative. We are developing a variety of strategies to resolve such discrepants. One of these is discussed below in the context of our direct test for Legionella bacteria.

We have examined a number of sputa which tested positive by the Legionella probe test and negative by culture. If the Legionella probe hybridizes significantly with a clinical sample, there are three possible interpretations.

1. The probe has bound in a sequence-dependent manner (true hybridization) to rRNA from a non-Legionella bacteria. In other words, the probe does not have the desired specificity.

2. The probe has bound in a sequence dependent manner to rRNA from a Legionella bacteria.

3. The probe has bound in a non-specific or non-sequence dependent manner to some component of the sample and subsequently behaves as if it is hybridized.

How do we distinguish among these possibilities?

There is no evidence that the Legionella genus probe will hybridize to non-Legionella bacteria when used under the conditions specified. The Legionella genus probe has been checked against rRNA's from a large number of different species and genera, including those found in the respiratory tract. The probe did not significantly hybridize with any of these rRNA's but did hybridize with all known Legionella species. Thus, the probe is specific only for Legionella species.

The physical chemical characteristics of DNA molecules make it possible to distinguish between sequence-specific binding (hybridization) and non-sequence specific binding to some sample component. All DNA probes are composed of the same 4 chemical sub-units, A, G, C and T. These bases are relatively simple chemical compounds and the base composition, as measured by G+C content, of the vast majority of DNA probes is similar. In general, therefore, the physical chemical characteristics of DNA probes are very similar. Consequently, if one probe will bind in a non-sequence dependent manner to a sample component, then any other probe will bind to that component in a non-sequence dependent manner. This suggests a simple method for determining whether non-sequence dependent binding has occurred in a sample which tests positive for Legionella genus probe and negative for culture. This is done by mixing the sample with a probe which has no target in the sample and then conducting the hybridization assay. If a "hybridization" signal

TABLE 11

MONOCLONAL ANTIBODY APPLICATIONS	AREAS OF OVERLAP	DNA PROBE APPLICATIONS
3 Immunoassay	3 Infectious Diseases	3 Genetic Diseases
3 Replacement	- Viruses	
	- Bacteria	
	3 Cancer	3 Disease Suscep- tibility
	3 Food Testing	3 Drug Resistance
	3 Plant Veterinary	
	3 Tissue Typing	
	3 Therapeutic	

TABLE 12

	GEN-PROBE DNA PROBE TEST	MCA TEST
Ease of Detection	Good to Excellent	Generally Excellent
Rapidity	Excellent	Excellent
Shelf Life of Probe or Antibody	Excellent	Excellent
Ability to Quantitate Organisms	Good to excellent	Not Good
Capability of Detecting 1-10 Bacterial Cells	Feasible	Not Feasible
Capability of Devel- oping Non-isotopic Tests	Excellent (on market)	Excellent (On Market)
Specificity	Essentially Absolute	Specificity is often a problem

is detected, then the probe has bound in a non- sequence dependent manner to a sample component. If no hybridization signal is seen, there is no non-sequence dependent binding of the probe occurring and Legionella target organisms are present in the sample. An example of such an analysis is discussed below.

We received a sputum sample which had tested positive with Legionella probe and negative with culture. After first confirming the probe-positive nature of the sample, a portion of the sample was hybridized with Mycobacteria genus probe. This probe did not significantly hybridize with the sample, thus ruling out non-sequence dependent binding of the Legionella probe and confirming the presence of Legionella in the sample.

TABLE 13

PROBE FORM	MCA FORM
1. Ideal probe is a single strand molecule which can only hybridize if target is present. 2. Probe can be either DNA or RNA.	1. MCA is a single molecule which can form Ag:Ab complex only if target antigen is present in the sample.

TABLE 14

PROBE SIZE	MCA SIZE
1. Length of probe can be as short as 10 bases (MW approx. 3.3×10^3) to greater than 10^4 bases (MW approx. 3.3×10^6).	1. IGG molecules have a MW approx. 1.5×10^5 equivalent to about 1500 amino acids.
2. The whole length of the probe is part of the combining site.	2. Size of combining site on antibody molecule is roughly 30 amino acids, a small fraction of the total.

DNA PROBE VERSUS MONOCLONAL ANTIBODY (MCA) TECHNOLOGY

These technologies overlap in certain areas and have unique roles in others (Table 11). DNA probe technology has only recently become commercialized while commercial antibody technology has been around for quite some time and such products are used widely in various areas. The main barrier to the widespread commercial use of DNA probes is the development of rapid, easy-to-use test formats. This is an area of intense investigation in the DNA probe industry and simple, easy-to-use formats will be forthcoming in the near future. Table 12 presents a summary of an overall comparison of the two technologies while Tables 13 thru 20 present detailed comparison of various aspects of probes and MCA's.

TABLE 15

SPECIFICITY OF PROBE	SPECIFICITY OF MCA
1. Depends upon:	1. Depends upon:
a) The base sequence of probe.	a) The primary, secondary and tertiary structure of the MCA.
b) The conditions under which the probe is used.	b) The primary, secondary and tertiary structure of the antigens.
2. Specificity requirements:	2. Specificty Requirements:
a) Probe must hybridize with target nucleic acid.	a) MCA must complex with target antigen.
b) Probe must not hybridize to other nucleic acids which may be present in the sample.	b) MCA must not complex with other antigens which may be present in the sample.
3. Specificity is essentially absolute.	3. MCA often complex with non-target antigens.

TABLE 16

PROBE PRODUCTION

PROBE TEST	MCA TEST
1. Probe sequence need not be known.	1. Ditto for MCA.
2. Sequence of target nucleic acid need not be known.	2. Ditto for MCA test.
3. Function of target sequence need not be known.	3. Ditto for MCA target.
4. Can select probe according to hybridization characteristic desired.	4. MCA selected to have desired specificty.
5. DNA probes can be developed very rapidly and cheaply. Gen-Probe has developed specific rationales for probe production which enable a probe to be identified and produced in 1-2 weeks.	5. MCA identification and production expensive and time-consuming. A short time for MCA development of an MCA is 6 months. The method used is basically a shotgun method with no rationale.

TABLE 17

RANGE OF REACTION CONDITIONS TOLERATED

PROBE TEST	MCA TEST
1. Nucleic acids are stable and will hybridize even in the presence of high concentrations of a wide variety of salts, organic solvent types and detergents, and over a wide range of temperatures.	1. The structure of the antigen and MCA are important to the association. Agents which greatly disturb the secondary and tertiary structure will make the antigens, the MCA, or both, incapable of associating. This greatly restricts the conditions used for the test.
2. Most DNA probe tests are done over a temperature range from room temperature to 72°C, and can be done over a range from -20°C to 100°C.	2. Most MCA tests are done over temperature range from room temperature to 37°C.

TABLE 18

"AVIDITY" OF PROBES

Probe + Target --> Probe:Target Hybrid	Antibody + Antigen<----->Complex
Once Probe:Target Hybrid is formed it does not spontaneously dissociate.	Ag:Ab Complex spontaneously dissociates. Rate of dissociation depends on avidity of Ab.

TABLE 19

RATE OF COMPLEX
FORMATION AND HYBRIDIZATION

HYBRIDIZATION	ANTIGEN ANTIBODY COMPLEX FORMATION
1. Standard methods for doing hybridization are much slower than comparable antigen-antibody reaction rates.	1. Very fast. Close to theo-retical for MCA-Hapten and within 10^2 - 10^3 of theoret-ical for MCA-large protein complexing. Much faster than standard hybridization methods.
2. The accelerated rate systems give rates comparable to the Ag:Ab rates.	

TABLE 20

ASSAY FOR PRESENCE
OF PROBE:TARGET HYBRIDS

HYBRIDIZATION	Ag + Ag COMPLEX FORMATION
1. Conventional methods are generally complex, lengthy and labor intense.	1. On the whole, the MCA tests are much easier and more rapid than conventional DNA probe tests.
2. Gen-Probe method is similar to many RIA type tests.	2. Generally comparable to the Gen-Probe test, although certain ones are faster and easier than the Gen-Probe tests.

CLINICAL SIGNIFICANCE OF DNA PROBES

DNA probes will allow the timely diagnosis of diseases caused by infectious agents. Early diagnosis and treatment will contribute greatly to both the patients' health and cost containment.

BIBLIOGRAPHY

1.   Britten, R. and D.E. Kohne. "Repeated Segments of DNA." Scientific American 1970. Vol. 222, No. 4, April 1970.

2.   "Nucleic Acid Hybridization: A Practical Approach." Ed. B.D. Hames and S.J. Higgins, IRL Press, Ltd., Oxford, Washington, D.C. 1985.

3.   "Current Communications in Molecular Biology, DNA Probes: Applications in Genetic and Infectious Disease and Cancer." Ed. L.S. Lerman. Cold Spring Harbor Laboratory. 1986.

4.   Brenner, D.J. "DNA Reassociation in the Taxonomy of Enteric Bacteria." Int. J. Systematic Bacteriology. Vol. 23, No. 4, pp. 298-307. 1973.

5.   Woese, C. "Archaebacteria." Scientific American. Vol. 244, No. 6, pp. 98. June 1981.

6.   Kohne D.E., et al. "Virus Detection by Nucleic Acid Hybridization. Examination of Normal and ALS Tissue for the Presence of Poliovirus." J. General Virology. Vol. 56, pp. 223-233. 1981.

THE INFLUENCE OF RAPID DIAGNOSIS OF STREPTOCOCCAL INFECTION ON PHARYNGITIS

AND RHEUMATIC FEVER

Joseph M. Campos

Children's Hospital National Medical Center
Washington, DC

INTRODUCTION

Pharyngitis is the most common of human infections. In the U.S. alone, an estimated 30-40 million physician visits each year are prompted by "sore throats". A recent survey of pediatricians revealed that throat cultures were performed during 14.4% of patient visits (14). Moreover, an incalculably high number of patients endure throat infection without ever seeking medical attention.

The impact of pharyngitis upon the American economy is staggering. Billions of dollars each year are spent on physician office visits and relief-providing medications. Employed adults, who stay home from work when ill, also stay home with infected children until they can return to schools or day-care centers. Substantial losses of productivity within the national workforce are the end result.

Group A streptococci are responsible for 30-40% of pharyngitis, with the rest attributable to an assortment of viruses. Pharyngitis is a self-limited infection. People who don't seek medical attention improve symptomatically within a few days. Nevertheless, the suppurative and nonsuppurative sequelae of Group A streptococcal pharyngitis can have very serious consequences and all cases of it should be treated with appropriate antimicrobial therapy.

Suppurative sequelae are rare and include otitis media, acute sinusitis, and tonsillar/retropharyngeal abscess. The latter complication, if left untreated, can be life-threatening. Of greater concern are the nonsuppurative sequelae, acute rheumatic fever and acute glomerulonephritis. These conditions may precipitate irreversible cardiac and renal damage. Antimicrobial therapy administered within ten days of infection effectively prevents acute rheumatic fever (23), but has no effect upon acute glomerulonephritis (72).

The incidence of acute rheumatic fever in the U.S. has been vanishingly small during the last two decades. However, sharply increased rates during the last three years were reported from Utah (67), Ohio (15,29), and Pennsylvania (70). In other areas of the world where overcrowded living conditions and difficult access to medical care are still commonplace, rheumatic fever has been, and continues to be, a major public health problem.

Rapid Methods in Clinical Microbiology
Edited by B. Kleger *et al.*
Plenum Press, New York

In the remainder of this chapter, I will discuss current options available to physicians for diagnosis of streptococcal pharyngitis, concentrating
upon recently developed assays for Group A streptococcal antigen. Performance of these assays in the physician's office and in the home will be
addressed. Lastly, I will speculate upon the coincidental rise in physician
use of antigen assays and the increased incidence of rheumatic fever in the
U.S.

DIAGNOSIS OF GROUP A STREPTOCOCCAL PHARYNGITIS

Many pediatricians and family practice physicians have performed cultures for Group A streptococci for years. In most settings, specimens are
inoculated to 5% sheep's bood agar, bacitracin disks are applied, and culture plates are incubated for 24 hours in an aerobic or candle-jar provided
atmosphere. Too often, the inexperience of the individuals reading cultures
and inadequacies of the culture method itself lead to under- or over-diag-
nosis of streptococcal pharyngitis.

The notion of a diagnostic test that rapidly and accurately identifies
patients who need antimicrobial therapy is very appealing to physicians.
Numerous assays have come on the market in recent years that furnish results
in less than ten minutes. Almost all of the products are immunoassays for
Group A streptococcal antigen which as been chemically or enzymatically
extracted from organisms enmeshed in throat swabs.

One product which is not an immunoassay detects PYR (L-pyrrolindonyl-
beta-naphthylamide) hydrolase, and enzyme produced by virtually all Group A
streptococci. Since this assay has fared quite poorly during clinical evaluations, (6,22), the remainder of the discussion of rapid tests will deal
only with the immunoassays.

ADVANTAGES OFFERED BY RAPID TESTS FOR STREPTOCOCCAL ANTIGEN

The advantages of rapid testing are summarized in Table 1. The most
conspicuous benefit is that infected patients may be diagnosed and treated
during the same visit. As mentioned previously, early therapy is not necessary for prevention of acute rheumatic fever. Nevertheless, if a patient's
antigen result is positive, antimicrobial treatment instituted during the
first visit theoretically is more effective in preventing the suppurative
sequelae of Group A streptococcal infection than the same treatment started
24-72 hours later.

Another benefit of first visit therapy was demonstrated by Lieu et al
(38), during a study in which duplicate throat swabs were collected from
patients who were presented with sore throats to the Emergency Department
of a large urban children's hospital. The first specimen was tested by
immunoassay shortly after collection. The second specimen was tested by
culture only if the immunoassay was negative. culture results were available after 24 hours. Antimicrobial treatment was withheld from patients
until a positive laboratory test result was available. Significantly
higher treatment rates were observed in children diagnosed by immunoassay
compared to culture (Table 2).

A number of studies have convincingly shown that prompt antimicrobial
therapy shortens the duration of streptococcal pharyngitis (33,47,52, 56).
The best designed study was a randomized, double-blind, prospective assessment of the effect of early penicillin therapy upon symptomatology. Statistically significant reductions in the duration of seven symptoms were
observed in patients treated during the first visit, compared to those in

Table 1. Advantages of Group A Streptococcal
Antigen Detection

1. Rapid Results
2. Early Administration of Antimicrobial Agents
a. Prevent sequelae of infection
b. Obtain greater compliance with therapy
c. Shorten the duration of illness
d. Shorten the period of communicability

Table 2. Patient Treatment Rates Based
On Manner of Diagnosis (38)

	Positive Result On	
	Rapid Test	Culture
Total Cases	96	93
Treated on 1st Visit	60	--
Treated on 2nd Visit	21	41
Never Treated	15	52
Treatment Rate	84%	44%

Table 3. Effect of Early Treatment on Symptoms
of Streptococcal Pharyngitis (52)

Beneficial Effect Upon:	Fever
	Sore Throat
	Dysphagia
	Lethargy
	Headache
	Abdominal Pain
	Anorexia
Detrimental Effect Upon:	Reinfection within four months

whom treatment was delayed for 48-56 hours (Table 3). Interestingly, the same study identified a singular drawback to early therapy. Patients treated on the day of presentation were more likely than patients treated later to experience subsequent infection with Group A streptococci in the succeeding four months. The authors hypothesized that early therapy blunted evolution of protective antibody in patients who suffered reinfection.

DISADVANTATES OF RAPID TESTS FOR STREPTOCOCCAL ANTIGEN

There are disadvantages to rapid testing that should not be overlooked (Table 4). The most serious is the lack of sensitivity of the commercially available assays compared to carefully performed cultures. Results of numerous evaluations conducted in my own laboratory reveal sensitivities ranging from 52.8% to 76.8% - figures certainly not high enough to justify replacing our culture method with any of these assays.

Results of a large number of studies evaluating more than half a dozen commercially available antigen assays have been published by my laboratory

Table 4.   Disadvantages of Group A
Streptococcal Antigen Detection

---

1.   Lack of Sensitivity of Currently Available Kits
2.   More Expensive than Culture
3.   Disruptive to Smooth Laboratory Operation

---

and others during the last four years (3,5,9,12,13,16,17,21,24,26,27,31,32,
38,39,41-44,48,50,54,55,58-63,65,68,69,73). Reported sensitivities fluctu-
ated between 44.9% and 100%. To add to the confusion, the same product
studied by different investigators frequently yields diverse results - a
perfect example being the Culturette[TM] Brand Ten-Minute Strep ID manufactured
by Marion Scientific, Kansas, MO (Table 5).

Several factors are responsible for disparity of results between investigators, including the sensitivity cf the culture method to which the antigen
was compared, the age of the study population, and the "blindedness" to the
study of the individuals collecting specimens and individuals performing
laboratory tests.

The specifics of the culture method greatly affect the study results:

1.   Nonselective versus selective blood agar
     The use of sheep blood agar containing antimicrobial agents permits
     growth of Group A streptococci and inhibits growth of other organisms
     greatly facilitates culture reading and increases the number of positive
     cultures detected (10,25,34,45,51).

2.   Aerobic versus anaerobic incubation of cultures
     Anaerobic incubation of culture plates enhances the readability and
     yield of positive culture (1,2,18,28,34,36,45,53,56) because strepto-
     cocci grow more luxuriantly in the absence of oxygen, and one of the
     beta-hemolysins of Group A streptococci, streptolysin O, is oxygen
     liable.

Table 5.   Sensitivity and Specificity of the Culturette[TM] Brand Ten-Minute
Strep ID as Reported by Various Investigators

Investigators	Sensitivity	Specificity	Reference
Berkowitz et al.	88%	98%	3
Bodino et al.	91%	97%	5
Campos and Charilaou	62%	99%	9
Chang and Mohla	90%	99%	12
Dubois et al.	78%	93%	17
Gerber et al.	83%	99%	22
Hampton et al.	100%	98%	27
Miceika et al.	92%	93%	42
Roddey et al.	72%	98%	59
Schwartz et al.	93%	90%	62
Slifkin and Gil	95%	100%	63
Strandjord et al.	74%	76%	65
Wagener and Remington	90%	96%	69
White et al.	78%	88%	73

3. 24 hours versus 48 hours of incubation
   The longer incubation period clearly yields a higher proportion of
   positive cultures (35-37,46).

4. Bacitracin versus immunoassay or PYR hydrolysis for identification of
   beta-hemolytic streptococci
   The bacitracin disk method identifies as many as 15% of non-Group A
   isolates as Group A streptococci. The immunoassay and PYR hydrolysis
   methods are considerably more accurate (6,7,11,19,40,64,71,74,75).

The more sensitive the culture method, the poorer the sensitivity exhibited
by the antigen assay. Misidentification of non-Group A streptococci as
Group A decreases the apparent specificity of the antigen assay.

Specimens obtained from children are likely to be of poorer quality
than those collected from adults because children are less cooperative in
allowing medical personnel to swab the posterior pharynx. Consequently,
specimens from children are frequently swabs of the tongue or palate instead
of the throat. Swabs from genuinely infected children may have such small
numbers of Group A streptococci that the antigen assay is negative despite
a 1+ or 2+ positive culture. Our own experiments revealed that several
assays require at least 1×10^4 to 8×10^5 colony forming units per swab
to yield positive results (J. M. Campos, C. Mohla, and S. Van Lierde, Abstr.
Third Eur. Cong. Clin. Microbiol. 1987, abstract 429). Analysis of age-
specific data from another of our studies confirmed that the combined sensitivity of two antigen assays was higher for specimens collected from
older than from younger children (Table 6) (9). Thus, product evaluations
involving specimens from young children may demonstrate lower sensitivities
than in adults.

Table 6. Effect of Patient Age Upon Sensitivity of
          Group A Streptococcal Antigen Detection Kits (9)

Patient Age	Sensitivity
Less than 14 years	61.4%
14 years or older	82.1%

Studies in which personnel collecting specimens are aware that an
evaluation is in progress tend to exaggerate the assay's capablility in
the "real world" setting. If the specimen collector is also a study invest-
igator, more than one specimen may be collected in order to obtain a better
specimen. Individuals who are not investigators will be more careful during
the collection process if it is known that the results of the specimen
testing will be carefully scrutinized. Likewise, if the individual performing the antigen test has prior knowledge of the culture findings, the reading of a borderline result could be influenced by that knowledge. The
most worthwhile evaluations are those in which the study design closely
mimics the manner in which the assay should be used in the physician's
office or microbiology laboratory.

Another notable disadvantage of rapid testing is the higher cost of
the assay compared to culture. List prices of kits range from $1.50 to
$6.00 per assay. The cost of a negative culture ranges from $0.25 to
$1.40 and that of a positive culture from $0.30 to $1.40. Culture costs
depend on use of nonselective versus selective blood agar and bacitracin

disk versus immunoassay or PYR hydrolysis for identification of beta-hemolytic streptococci. The overall cost of testing strikingly escalates if the oft-recommended practice of culturally confirming negative antigen results is adopted. Assuming a culture positive rate of 25%, the estimate for costs of materials alone would increase by a minimum of 82% per specimen (8). A comparable increase in workload per specimen would be expected also.

One last disadvantage of rapid testing is disruption of the workflow of a busy office or laboratory. To be most useful, test results should be available promptly so that antimicrobial treatment, if needed, can be administered before the patient leaves for home. Few office laboratories and even fewer microbiology laboratories are staffed to carry out "stat" testing on more than an occasional specimen. Optimum use of the rapid assay requires that a trained individual be available for testing whenever a specimen is collected. Staff expansion or restructuring of employee responsibilities may be necessary to achieve that ideal.

FALSE POSITIVE AND FALSE NEGATIVE ANTIGEN TEST RESULTS

No laboratory test is perfect. False positive and false negative results always occur. Compared to culture, most evaluations of assays for streptococcal antigen have shown a much lower incidence of false positive than false negative results. Possible reasons for false positive results are listed in Table 7.

In these studies, crossreactions with non Group A streptococcal antigens has not been a serious problem. With regard to culturing, if the reference culture method is so poorly conceived that the antigen assay is more sensitive than the "gold standard", then truly positive antigen results will appear to be falsely positive. Nonhemolytic Group A streptococci very likely would be missed by standard culture methods even when present in large enough numbers to yield positive antigen results, because they are uncommon causes of pharyngitis (30). Large numbers of nonviable Group A streptococci in specimens that yield positive antigen and negative culture results was reported in one study (K. P. Aspden, Abstr. Annu. Meet. Am. Soc. Microbiol. 1984, C210, p. 270). The authors concluded that patients who had recently ingested a popular over-the-counter cold remedy experienced transient reductions in viable pharyngeal Group A streptococci without diminishing the antigen load.

The potential causes of false negative antigen results shown in Table 8 were discussed earlier in the chapter. The lack of assay sensitivity observed during our own evaluations has meant that specimens yielding 1+ (10 colonies) and 2+ (11-100 colonies) Group A streptococci by culture usually give rise to negative antigen results. Some have questioned the clinical significance of 1+ and 2+ cultures, and implied that failure to detect such specimens is of no consequence. That logic is fallacious in that it assumes all throat swab specimens are properly collected. That

Table 7. Reasons for False Positive Group A
Streptococcal Antigen Detection Results

---

1. Lack of Kit Specificity
2. Nonviable Group A Streptococci in Specimens
3. Nonhemolytic Group A Streptococci in Specimens
4. Culture Method less Sensitive than Antigen Assay

Table 8. Reasons for False Negative Group A
Streptococcal Antigen Detection Results

---

1. Lack of Kit Sensitivity
2. Poorly Collected Throat Swabs
3. Inaccurate Identification of Culture Isolates

---

assumption certainly does not hold true when children are tested, as stated
earlier in the chapter.

Gerber et al. reported data from a recent study which supports this
contention (20). Throat swabs from children were tested by culture and
antigen assay, and acute/convalescent sera were tested for streptolysin O
and DNAse B antibodies. The percentage of patients demonstrating a four-
fold or greater rise in either antibody titer was not significantly different in the patients who were culture positive/antigen positive and culture
positive/antigen negative (Table 9). Since the latter group contained
almost all of the 1+ culture positive patients, the rate of true infection
in that subpopulation was the same as in the 2+ - 4+ culture positive
patients.

IMPACT OF STREPTOCOCCAL ANTIGEN ASSAYS UPON PATIENT MANAGEMENT

Prior to the availability of kits for detection of streptococcal antigen, physicians had several options for management of patients with pharyngitis:
1. Initiate Antimicrobial therapy in all patients
2. Order throat culture and initiate antimicrobial therapy in all
    patients
3. Order throat culture, initiate partial antimicrobial therapy
    in all patients and refill prescription only if culture is
    positive
4. Order throat culture and initiate antimicrobial therapy only
    if culture is positive

The first option is unsound in that indiscriminate administration of
antimicrobial therapy is a risky practice. In fact, projections estimate
that if everyone with a sore throat were given an antimicrobial agent,
approximately forty times as many patients would suffer potentially life-
threatening allergic reactions than would be prevented from progressing to
acute rheumatic fever (49).

The second option is even more outlandish in that the concern over the
first option still applies, and the patient, the patient's insuance company
or society would have to pay for culture results that have no influence
at all upon the ultimate course of management.

Table 9. Group A Streptococcal Antigen Dectection Results Versus
Streptococcal Antibody Titer Rise (20)

Test Result	Number	Titer Rise
Culture +, Antigen +	224	114 (51%)
Culture +, Antigen -	31	14 (45%)

The third option is founded upon better logic, but still risks unnecessary allergic reactions to antimicrobial angents. Furthermore, patients who improve symptomatically after initiation of treatment may not feel compelled to refill the prescription. Less than a full course of antimicrobial therapy may not prevent rheumatic fever.

The last option is the most defensible, but has negative aspects as well. Patients who clinically improve during the 24-72 hours that positive cultures are incubating may elect to ignore their physician's advice and not fill the prescription. Also, patients who are ill enough to seek medical attention may be unhappy returning home without a prescription or a diagnosis. That same patient may seek care from a different physician when other medical problems occur in the future.

The existence of rapid assays for Group A streptococcal antigen has added to the management options listed above. The clinical merits of this testing cannot be realistically assessed until the effects of test results upon physician decision-making and patient management are actually gauged.

A positive impact of rapid testing was documented in a retrospective study published by True and colleagues (Table 10) (66). The study was conducted in the family practice model office at the University of Iowa College of Medicine. During Year 1, a year when rapid testing was not performed, most patients were treated with an antimicrobial agent before culture results were available. During Year 2, a year when rapid testing was performed, the majority of patients were treated only after test results were available. Significantly fewer episodes of inappropriate antimicrobial therapy were observed after routine institution of rapid testing.

A more recently published study surveyed pediatricians by questionnaire about their attitudes toward rapid testing (4). For a hypothetical assay with a sensitivity of 82%, a specificity of 98%, results available in 20 minutes, and a likelihood of infection of 0.58, 84% of the pediatricians stated they would order such a test. Overall, the authors forecasted a 54% reduction in the number of cultures ordered. They predicted the percentage of patients receiving antimicrobial therapy would be reduced from 86% to 65%. Only 54% of the respondents indicated they would order the rapid test if the likelihood of infection was 0.09. Accordingly, the projected reductions in the number of cultures ordered and percentage of patients receiving antimicrobial therapy were less pronounced.

Table 10. Effect of Group A Streptococcal Antigen Detection Test Results Upon Patient Management (66)

	Management Option	Year 1	Year 2
1.	Symptomatic Treatment Only	58%	63%
2.	Antimicrobial Agent Prescribed Before Test Result Available	27%	9%
3.	Antimicrobial Agent Prescribed After Test Result Available	10%	26%
4.	Antimicrobial Agent Prescription Filled Only if Culture Positive	6%	2%

The full impact of the antigen assays upon clinical practice is yet to be felt. Only 20% of the pediatricians queried by Berwick et al. had ever used an antigen assay in their offices (4). During an informal survey of my own, a large majority of hospital microbiology laboratories still rely upon culture for definitive diagnosis of streptococcal pharyngitis.

The economic advantage accrued by physicians who conduct their own diagnostic testing is substantial. Current regulations permit physicians to charge "what the market will bear" for tests performed on the premises, whereas only a small processing fee can be added to the cost of tests referred to outside laboratories. As assay procedures become simpler, so that little-trained or untrained individuals can perform assays successfully, the volume of testing carried out in the physician's office will soar. In anticipation, the marketing strategies of test kit manufacturers already are targeting private practice physicians.

GROUP A STREPTOCOCCAL ANTIGEN TESTING IN THE HOME

Several manufacturers are considering development of Group A streptococcal antigen assays for use in the home. It is improbable that these ventures will be successful for the following reasons:

1. The difficulty and discomfort entailed in collection of specimens may deter most of the general public from purchasing test kits.
2. Even if a home test result is positive, the physician will insist upon repeating the test before prescribing antimicrobial therapy. This will discourage home testing.
3. The potential liability for manufacturers when ill-performed tests yield inaccurate results may in the end dissuade most manufacturers from pursuing the home test market.

STREPTOCOCCAL ANTIGEN TESTING AND ACUTE·RHEUMATIC FEVER

Finally, a comment upon the possible association between the recent upsurge in acute rheumatic fever in the U.S. and increasing physician dependence upon rapid antigen assays for diagnosis of Group A streptococcal pharyngitis. I am not aware of any data that document such an association exists. Since a prospective study design would be unethical, evidence can be obtained only from a retrospective review of physician practices in the affected areas. The coincidental timing of the two phenomena is suggestive and merits further investigation. One hopes that these investigations are already in progress.

CONCLUSION

During the last 40 years, a great deal has been written about diagnosis of Group A streptococcal pharyngitis. Refinements in cultural technique dominated the literature through the 1970's. The 1980's heralded development of noncultural methods that furnish results in minutes rather than days. Without a doubt, culture as a diagnostic tool for streptococcal pharyngitis one day will be of historical interest only.

Physicians and laboratories must resist the temptation to depend solely upon these assays prematurely. A number of carefully conducted evaluations have demonstrated an inherent lack of sensitivity in the available immunoassays. Undiagnosed infection can have devastating consequences for the patient. Until an assay is developed with a sensitivity comparable to that of a well-performed culture, I recommend that physicians and laboratorians

not rely upon negative antigen assay results. Antigen positive patients
should be treated – antigen negative patients whould be cultured and treated
only if the culture is positive.

REFERENCES

1. Beerman, C. A., and S. A. Goldblatt, Screening for Group A strepto-
        coccus by means of anaerobic primary plate technique, J. Pediatr.
        101:70–72 (1982).
2. Belli, D. C., R. Auckenthaler, L. Paunier, and P. D. Ferrier, Throat
        cultures for Group A beta–hemolytic streptococcus, Am. J. Dis.
        Child. 138:274–476 (1984).
3. Berkowitz, C. D., B. F. Anthony, E. L. Kaplan, E. Wolinsky, and A. L.
        Bisno, Cooperative study of latex agglutination to identify Group
        A streptococcal antigen on throat swabs in patients with acute
        pharyngitis, J. Pediatr. 107:89–92 (1985).
4. Berwick, D. M., E. Gorss, A. B. Macone, E. J. O'Rourke, and D. A.
        Goldmann, Impact of rapid antigen tests for Group A streptococcal
        pharyngitis on physician use of antibiotics and throat cultures.
        Pediatr. infect. Dis. J. 6:1095–1102 (1987).
5. Bodino, J. A., E. L. Lopez, E. Rubeglio, And G. G. de Giavedoni,
        Evaluation of a rapid test for Group A Streptococcus at a
        physician's office and hospital laboratory in Buenos Aires,
        Argentina, Pediatr. Infect. Dis. J. 6:762–764 (1987).
6. Bracker, M. D., and N. J. Lugo, Evaluation of the fluorescent test
        for office–based detection of Group A beta–hemolytic streptococcal
        infection, J. Fam. Pract. 23:439–441 (1986).
7. Burdash, N. M., M. E. West, R.T. Newell, and G. Teti, Group identifi-
        cation of streptococci: evaluation of three rapid agglutination
        methods, Am. J. Clin. Pathol. 76:819–822 (1981).
8. Campos, J. M., Noncultural diagnosis of Group A streptococcal pharyn-
        gitis, Clin. Microbiol. Newsl. 9:152–154 (1985).
9. Campos, J. M., and C. C. Charliaou, Evaluation of Detect-A-Strep and
        the Culturette Ten–Minute Strep ID kits for detection of Group A
        streptococcal antigen in oropharyngeal swabs from children, J.
        Clin. Microbiol. 22:145–148 (1985).
10. Carlson, J.R., W. G. Mertz, B. E. Hansen, S. Ruth, and D. G. Moore,
        Improved recovery of Group A beta–hemolytic streptococci with a
        new selective medium, J. Clin. Microbiol. 21:307–309 (1985).
11. Castle, D., S. Dessock-Philip, and C. S. F. Eason, Evaluation of an
        improved Streptex kit for the grouping of beta–hemolytic strepto-
        cocci by agglutination, J. Clin. Pathol. 35:719–722 (1982).
12. Chang, M. F., and C. Mohla, Ten–minute detection of Group A strepto-
        cocci in pediatric throat swabs, J. Clin. Microbiol. 21:258–259
        (1985).
13. Clegg, H. W., O. F. Roddey, Jr., C. U. Manuey, R. L. Swetenburg, and
        E. S. Martin, Rapid diagnosis of streptococcal pharyngitis using
        enzyme immunoassay, Pediatr. Infect. Dis. J. 6:696–697 (1987).
14. Colchi, S. L., Diagnosis and treatment of streptococcal pharyngitis:
        survey of U.S. medical practitioners, in: "Pharyngitis: Manage-
        ment in an Era of Declining Rheumatic Fever," S. T. Shulman, ed.,
        Prager, New York, NY, (1984), p. 23–31.
15. Cogeni, B., C. Rizzo, J. Congeni, and V. V. Screenivasan, Outbreak of
        acute rheumatic fever in northeasy Ohio, J. Pediatr. 111:176–179
        (1987).
16. Dobkin, D., and S. T. Shulman, Evaluation of an ELISA for Group A
        streptococcal antigen for diagnosis of pharyngitis, J. Pediatr.
        110:566–569 (1987).

17. Dubois, D., V. G. Ray, B. Nelson, and J. B. Peacock, Rapid diagnosis of Group A strep pharyngitis in the emergency department, _Ann. Emer. Med._ 15:157–159 (1986).

18. Dykstra, M. A., J. C. McLaughlin, and R. C. Bartlett, Comparison of media and techniques for detection of Group A streptococci in throat swab specimens, _J. Clin. Microbiol._ 9:236–238 (1979).

19. Facklam, R. R., L. G. Thacker, B. Fox, and L. Eriquez, Presumptive identification of streptococci with a new test system, _J. Clin. Microbiol._ 15:987–990 (1982).

20. Gerber, M. A., M. F. Randolph, J. Chanatry, L. L. Wright, K. K. DeMeo, and L. R. Anderson, Antigen detection test for streptococcal pharyngitis: evaluation of sensitivity with respect to true infections, _J. Pediatr._ 108:654–658 (1986).

21. Gerber, M. A., L. J. Spadaccini, L. L. Wright, and L. Deutsch, Latex agglutination tests for rapid identification of Group A streptococci directly from throat swabs, _J. Pediatr._ 105:702–705 (1984).

22. Gerber, M. A., M. F. Randolph, and R. C. Tilton, Enzyme fluorescence procedure for rapid diagnosis of streptococcal pharyngitis, _J. Pediatr._ 108:421–423 (1986).

23. Gordis, L., Effectiveness of comprehensive care programs in preventing rheumatic fever, _N. Engl. J. Med._ 289:331–335 (1973).

24. Granato, P. A., L. M. Giachetti-Powers, K. S. Murfitt, and D. King, Comparative evaluation of PathoDx Strep A Test and culture for the detection of Group A streptococci in pharyngeal specimens, _Diagn. Microbiol. Infect. Dis._ 5:293–298 (1986).

25. Gunn, B. A., D. K. Ohashi, C. A. Gaydos, and E. S. Holt, Selective and enhanced recovery of Group A and B streptococci from throat cultures with sheep blood agar containing sulfamethoxazole and trimethoprim, _J. Clin. Microbiol._ 5:650–655 (1977).

26. Hadfield, S. G., D. N. Petts, P. Kennedy, A. Lane, and M. B. McIllmurray, Novel color test for rapid detection of Group A streptococci, _J. Clin. Microbiol._ 25:1151–1154 (1987).

27. Hampton, K. D., B. L. Wasilauskas, and H. W. Johnson, Detection of Group A streptococci from throat cultures: comparison of enzyme fluorescence, latex agglutination, and routine culture, _Lab. Med._ 18:37–38 (1987).

28. Hayden, G. F., S. Dudley, and J. O. Hendley, Use of an anaerobic culture jar in processing pediatric throat cultures, _Clin. Pediatr._ 23:224–227 (1984).

29. Hosier, D. M., J. M. Craensen, D. W. Teske, and J. J. Wheller, Resurgence of acute rheumatic fever, _Am. J. Dis. Child._ 141:730–733 (1987).

30. James, L., and R. B. McFarland, An epidemic of pharyngitis due to a non-hemolytic Group A streptococcus at Lowry Air Force Base, _N. Engl. J. Med._ 284:750–752 (1971).

31. Kellogg, J. A., and J. P. Manzella, Detection of Group A streptococci in the laboratory or physician's office, _J. Amer. Med. Assoc._ 255:2638–2642 (1986).

32. Kellogg, J. A., R. C. Landis, A. S. Nussbaum, and D. A. Bankert, Performance of an enzyme immunoassay test and anaerobic culture for detection of Group A streptococci in a pediatric practice versus a hospital laboratory, _J. Pediatr._ 111:18–21 (1985).

33. Krober, M. S., J. W. Bass, and G. N. Michels, Streptococcal pharyngitis: placebo-controlled double-blind evaluation of clinical response to penicillin therapy, _J. Am. Med. Assoc._ 253:1271–1274 (1985).

34. Kurzynski, T., and C. Meise Van Holton, Evaluation of techniques for isolation of Group A streptococci from throat cultures, _J. Clin. Microbiol._ 13:891–894 (1981).

35. Kurzynski, T. A., and C. K. Meise, Evaluation of sulfamethoxazole-
    trimethoprim blood agar plates for recovery of Group A strepto-
    cocci from throat cultures, J. Clin. Microbiol. 9:189–193 (1979).
36. Lauer, B. A., L. B. Reller, and S. Mirrett, Effect of atmosphere and
    duration of incubation on primary isolation of Group A strepto-
    cocci from throat cultures, J. Clin. Microbiol. 17:338–340 (1983).
37. Libertin, C. R., A. D. Wold, and J. A. Washington II, Effects of
    trimethoprim-sulfamethoxazole and incubation atmosphere on iso-
    lation of Group A streptococci, J. Clin. Microbiol. 18:680–682
    (1983).
38. Lieu, T. A., G. R. Fleisher, and J. S. Schwartz, Clinical performance
    and effect on treatment rates of latex agglutination testing for
    streptococcal pharyngitis in an emergency department, Pediatr.
    Infect. Dis. 5:655–659 (1986).
39. Matteson, M. L., and J. P. Anhalt, Effect of delay in processing on
    the performance of Directigen for the detection of Group A
    streptococci in throat swabs, J. Clin. Microbiol. 21:993–994
    (1986).
40. Matthieu, D. C., B. L. Wasilauskas, and R. A. Stallings, A rapid
    staphylococcal coagglutination technic to differentiate Group A
    from other streptococcal groups, Am. J. Clin. Pathol. 72:463–467
    (1979).
41. McCusker, J. J., E. L. McCoy, C. L. Young, R. Alamares, and L. S.
    Hirsch, Comparison of Directigen Group A Strep Test with a
    traditional culture technique for detection of Group A beta-
    hemolytic streptococci, J. Clin. Microbiol. 20:824–825 (1984).
42. Miceika, B. G., A. S. Vitous, and K. D. Thompson, Detection of Group
    A streptococcal antigen directly from throat swabs with a ten-
    minute latex agglutination test, J. Clin. Microbiol. 21:467–469
    (1985).
43. Miller, J. M., H. L. Phillips, R. K. Graves, and R. R. Facklam,
    Evaluation of the Directigen Group A Strep Test kit, J. Clin.
    Microbiol. 20:846–848 (1984).
44. Mills, E. L., M. I. Hunter, D. W. Scheifele, R. A. Bortolussi, R.
    Gold, N. E. MacDonald, and M. Rola-Pleszczynski, Rapid identi-
    fication of Group A beta-hemolytic streptococci in throat swabs,
    Can. Med. Assoc. J. 134:228–229 (1986).
45. Mirrett, S., J. S. Monahan, and L. B. Reller, Comparative evaluation
    of medium and atmosphere of incubation for isolation of
    Streptococcus pyogenes, Diagn. Microbiol. Infect. Dis. 6:217–221
    (1987).
46. Murray, P. R., A. D. Wold, C. A. Schreck, and J. A. Washington II,
    Effects of selective media and atmosphere of incubation on the
    isolation of Group A streptococci, J. Clin. Microbiol. 4:54–56
    (1976).
47. Nelson, J. D., The effect of penicillin therapy on the symptoms and
    signs of streptococcal pharyngitis, Pediatr. Infect. Dis. 3:10–13
    (1984).
48. Ogay, K., and J. Bille, Rapid coagglutination test for the direct
    detection of Group A streptococci from throat swabs, Eur. J. Clin.
    Microbiol. 5:317–319 (1986).
49. Pantell, R. H., Cost-effectiveness of pharyngitis management and
    prevention of rheumatic fever, Ann. Intern. Med. 86:497–499 (1977).
50. Patel, K., A. L. Chittom, R. Toshniwal, and F. E. Kocka, Rapid
    commercial test for direct detection of Group A streptococci in
    throat swabs, Eur. J. Clin. Microbiol. 6:193–194 (1987).
51. Petts, D. N., Colistin-oxolinic acid-blood agar: a new selective
    medium for streptococci, J. Clin. Microbiol. 19:4–7 (1984).
52. Pichichero, M. E., F. A. Disney, W. B. Talpey, J. L. Green, A. B.
    Francis, K. J. Roghmann, and R. A. Hoekelman, Adverse and bene-
    ficial effects of immediate treatment of Group A beta-hemolytic

streptococcal pharyngitis with penicillin, Pediatr. Infect. Dis. J. 6:635-643 (1987).

53. Pien, F. C., C. L. Ow, N. S. Isaacson, N. T. Goto, and R. C. Rudoy, Evaluation of anaerobic incubation for recovery of Group A streptococci from throat cultures, J. Clin. Microbiol. 10:392-393 (1979).

54. Radetsky, M., J. A. Solomon, and J. K. Todd, Identification of streptococcal pharyngitis in the office laboratory: reassessment of new technology, Pediatr. Infect. Dis. J. 6:556-563 (1987).

55. Radetsky, M., R. C. Wheeler, M. H. Roe, and J. K. Todd, Comparative evaluation of kits for rapid diagnosis of Group A streptococcal disease, Pediatr. Infect. Dis. J. 4:274-281 (1985).

56. Randolph, M. F., M. A. Gerber, K. K. DeMeo, and L. Wright, Effect of antibiotic therapy on the clinical course of streptococcal pharyngitis, J. Pediatr. 106:870-875 (1985).

57. Randolph, M. F., J. J. Redys, and J. B. Cope, Evaluation of aerobic and anaerobic methods of recovery of streptococci from throat cultures, J. Pediatr. 104:897-899 (1984).

58. Reichwein, B., D. Jungkind, M. Guardiani, R. Gilbert, G. Prosswimmer, and P. Amadio, Comparison of two rapid latex agglutination methods for detection of Group A streptococcal pharyngitis, Am. J. Clin. Pathol. 86:529-532 (1986).

59. Roddey, O. F., H. W. Clegg, L. T. Clardy, E. S. Martin, and R. L. Swetenburg, Comparison of a latex agglutination test and four culture methods for identification of Group A streptococci in a pediatric office laboratory, J. Pediatr. 108:347-351 (1986).

60. Rudin, L., J. Rotta, C. Blomquist, A. Benslimane, O. Berger-Jekic, T. Kereselidze, K. Prakash, A. Sukonthaman, L. Tay, and E. Tikhomirov, Multicentre evaluation of a direct coagglutination test for Group A streptococci, Eur. J. Clin. Microbiol. 6:303-305 (1987).

61. Schwabe, L. D., M. T. Small, and E. L. Randall, Comparison of TestPack Strep A Test kit with culture technique for detection of Group A streptococci, J. Clin. Microbiol. 25:309-311 (1987).

62. Schwartz, R. H., G. F. Hayden, P. McCoy, T. Sait, and O. M. Chhabra, Rapid diagnosis of streptococcal pharyngitis in two pediatric offices using a latex agglutination kit, Pediatr. Infect. Dis. J. 4:647-650 (1985).

63. Slifkin, M., and G. M. Gil, Evaluation of the Culturette Brand Ten-Minute Group A Strep ID technique, J. Clin. Microbiol. 20:12-14 (1984).

64. Slifkin, M., and G. R. Pouchet-Melvin, Evaluation of three commercially available test products for serogrouping beta-hemolytic streptococci, J. Clin. Microbiol. 11:249-255 (1980).

65. Strandjord, T. P., E. J. Rich, and L. Quan, Comparison of two antigen detection techniques for Group A streptococcal pharyngitis in a pediatric emergency department, Pediatr. Infect. Dis. J. 6:1071-1072 (1987).

66. True, B. L., B. L. Carter, C. E. Driscoll, and J. D. House, Effect of a rapid diagnostic method on prescribing patterns and ordering of throat cultures for streptococcal pharyngitis, J. Fam. Pract. 23:215-219 (1986).

67. Veasy, L. G., S. E. Wiedmeier, G. S. Orsmond, H. D. Ruttenberg, M. M. Boucek, S. J. Roth, V. F. Tait, J. A. Thompson, J. A. Daly, E. L. Kaplan, and H. R. Hill, Resurgence of acute rheumatic fever in the intermountain region of the United States, N. Engl. J. Med. 316:421-427 (1987).

68. Venezia, R. A., A. Ryan, S. Alward, and W. A. Kostun, Evaluation of a rapid method for the detection of streptococcal Group A antigen directly from throat swabs, J. Clin. Microbiol. 21:395-398 (1985).

69. Wagener, S., and J. S. Remington, Rapid diagnosis of streptococcal
        pharyngitis, J. Pediatr. 107:155-156 (1985).
70. Wald, E. R., B. Dashefsky, C. Feidt, D. Chiponis, and C. Byers, Acute
        rheumatic fever in western Pennsylvania and the tristate area,
        Pediatrics 80:371-374 (1987).
71. Wasilauskas, B. L., and K. D. Hampton, Evaluation of the Strep-A-Fluor
        identification method for Group A streptococci, J. Clin. Microbiol.
        20:1205-1206 (1984).
72. Weinstein, L., and J. LeFrock, Does antimicrobial therapy of strepto-
        coccal pharyngitis or pyoderma alter the risk of glomerulneph-
        ritis?, J. Infect. Dis. 124:229-231 (1971).
73. White, C. B., J. W. Bass, and S. M. Yamada, Rapid latex agglutination
        compared with the troat culture for the detection of Group A
        streptococcal infection, Pediatr. Infect. Dis. 5:208-212 (1986).
74. Wu, T. C., E. C. Williams, and P. S. Conville, Rapid identification of
        Group A streptococci by the Strep-A-Fluor system, Diagn. Microbiol.
        Infect. Dis. 6:5-9 (1987).
75. Yajko, D. M., J. Lawrence, P. Nassos, J. Young, and W. K. Hadley,
        Clinical trial comparing bacitracin with Strep-A-Chek for accuracy
        and turnaround time in the presumptive identification of
        Streptococcus pyogenes, J. Clin. Microbiol. 24:431-434 (1986).

THE USE OF DNA PROBES FOR RAPIDLY IDENTIFYING CULTURES OF MYCOBACTERIUM

Charles L. Woodley, Vella A. Silcox,
Margaret M. Floyd and George P. Kubica

Mycobacteriology Laboratory
Division of Bacterial Diseases
Center for Infectious Diseases
Centers for Disease Control
Atlanta, Georgia 30333

INTRODUCTION

There has been an explosion of new technology for the rapid, specific identification of many microorganisms pathogenic for humans. These new methods have been especially welcomed in the mycobacteriology laboratory, plagued by long bacterial generation times that translated into specific identification periods measured in months. The race to develop these methods and to make them easy to perform and the secrecy surrounding new patent applications for some of the unique procedures employed have spawned such terms as "dipstick- or black box-technology" to describe these welcome additions to diagnostic microbiology.

Single-stranded deoxyribonucleic acid (DNA) probes developed and sold by Gen-Probe, Inc., San Diego, CA., are an example of these new, rapid procedures. Probes specific for Legionella species[4,8] and Mycoplasma pneumoniae[7] were first on the scene. These were followed closely by 4 probes designed for confirmation of pure cultures at various levels of identification within the taxon Mycobacterium: the genus-specific probe, developed to identify all species within this genus; the Mycobacterium tuberculosis complex probe that bound to the various species or biovars within this complex (i.e., M. tuberculosis, M. bovis, BCG strains of M. bovis, M. microti, and M. africanum); and, finally, 2 probes complementary to the ribosomal ribonucleic acids (rRNA) of M. avium and M. intracellulare.

Dr. Kohne has already described (earlier, this book) the rationale and advantages of developing a DNA probe system targetted for in-solution detection of rRNA. I want to share with you our experiences in some of the initial investigations of these probes, as well as our continuing evaluations of these products in our daily reference laboratory activities at the Centers for Disease Control (CDC).

MATERIALS AND METHODS

We evaluated the 4 commercially available (Gen-Probe, Inc.), single-stranded, [125] I-labeled DNA mycobacterial probe kits designed for culture confirmation in the sequence in which they were produced.

Rapid Methods in Clinical Microbiology
Edited by B. Kleger *et al.*
Plenum Press, New York

All cultures used were obtained from the Trudeau Mycobacterial Culture
Collection, the American Type Culture Collection, or from carefully
studied reference cultures identified by the Mycobacteriology
Laboratory, Center for Infectious Diseases, CDC.

We tested the genus-specific probe with 200 strains of
microorganisms: 185 cultures representing strains of 27 different
species of Mycobacterium and 15 strains of 8 other genera.

The M. tuberculosis complex probe was tested with 240 strains: 128
strains in the M. tuberculosis complex, 106 strains representing 22
other species of Mycobacterium, and 1 strain each of Actinomadura
madurae, Nocardia brasiliensis, N. otitidiscaviarum, Rhodococcus equi,
R. rubropertinctus and R. terrae.

The Mycobacterium avium Complex Culture Confirmation Kit,
containing probes specific for both M. avium and M. intracellulare, was
evaluated with 117 strains of mycobacteria: 60 strains initially assumed
to be M. avium complex on the basis of biochemical tests and serotyping
(37 were M. avium and 23 were M. intracellulare) and 57 strains of other
mycobacteria. When serotyping cultures in the M. avium complex (MAC),
serovars 1 through 6 and 8 through 11 are currently recognized as M.
avium, while serovars 7 and 12 through 19 are accepted as M.
intracellulare. Serovars 20 to 28 need more careful study before
species names are ascribed to them.

In all cases the observed probe test results (i.e., positive or
negative) were compared with expected results, as derived from
biochemical data[6] serotyping[5] high performance liquid chromatography
(HPLC) analysis of mycolic acids[1-3], and DNA-DNA sequence similarity
performed by Gen-Probe personnel on coded cultures.

Performance of probe tests. Because many details pertaining to
probe production and test performance were (and some still are)
proprietary, relatively little precise information can even now be
provided. The procedure initially used to evaluate the probes varied
slightly, though not substantively, from that now described in the
package inserts of the various probes. For each run, known
probe-positive and -negative cultures were always included. For each
culture tested, the ratio of counts from pelletized, hybridized probe to
the starting unhybridized probe (i.e., total count) was calculated as a
percentage. When percentage hybridization was ≥ 10% for the genus probe
or any of the species probes, identification was considered to have been
made.

Identification as
Mycobacterium

		+	−
Gen-Probe	+	184	0
Test Results	−	1	15

Sensitivity: 184/185 = 99.5%
Specificity: 15/15 = 100%

Figure 1. Evaluation of the Mycobacterium genus-
specific probe (Gen-Probe) with 200
cultures identified by CDC (see text
for details)

```
                    Identification as
                 M. tuberculosis complex

                         +      -

        Gen-Probe    +    128     1

        Test Results  -    0     111

        Sensitivity:  128/128  = 100%
        Specificity:  111/112  =  99.1%
```

Figure 2. Evaluation of the M. tuberculosis complex-
         specific probe (Gen-Probe) with 240
         cultures of Mycobacterium and closely-
         related genera (see text for details)

RESULTS

Genus-specific probe. We examined 200 cultures with the
genus-specific probe. Only 1 strain of 185 Mycobacterium cultures
failed to give a positive probe result (thus, test sensitivity was
99.5%), and none of the other genera reacted with the probe (specificity
was 100%)(Fig. 1). The 1 culture of Mycobacterium that failed to react
was a strain of M. terrae complex, a saprophytic environmental organism.

M. tuberculosis complex probe. All 128 strains of M. tuberculosis
complex tested hybridized with the probe (sensitivity was 100%), while
only 1 strain other than M. tuberculosis reacted with the probe
(specificity 99.1%) (Fig. 2). Just as in the genus probe studies, the 1
aberrant strain, though still unidentified, was an M. terrae-like
organism (though different from the M. terrae in the genus probe study).

M. avium complex probe kit. In view of our successful results
with both the genus- and M. tuberculosis complex-specific probes, our
initial results with the M. avium (Fig. 3) and M. intracellulare (Fig.
4) probes were disheartening; test sensitivities were only 84% and
74%, respectively. Eight strains of M. avium (Fig. 3) and 9 strains
of M. intracellulare (Fig. 4), yielded unexpected probe results;
because several of the seemingly aberrant strains were examined by
both probes, the actual number of strains labeled as "incorrect" was
only 13 (11%). These 13, together with 6 other strains that had given
initially "correct" results, were reexamined in much greater detail to
be certain of their precise identity; cultures with initially
"correct" results were included to be certain that the simple act of

```
                  Initial identification as
                M. avium, using biochemical
                    and serologic tests

                         +      -

        Gen-Probe    +    31      2

        Test Results  -    6      78

        Sensitivity:  31/37 = 84%
        Specificity:  78/80 = 98%
```

Figure 3. Evaluation of the M. avium-specific probe
         (Gen-Probe) with 60 suspected M. avium complex
         isolates and 57 other mycobacteria (see text).

```
                    Initial identification as
                    M. intracellulare, using
                    biochemical and serologic
                    tests

                                 +    -

        Gen-Probe       +      17    3

        Test Results    -       6   91

        Sensitivity:  17/23 = 74%
        Specificity:  91/94 = 97%
```

Figure 4. Evaluation of the M. intracellulare-specific
           probe (Gen-Probe) with 60 suspected M. avium
           complex isolates and 57 other mycobacteria
           (see text)

retesting would not alter previously observed results. Repeated tests
included differential biochemical tests, probe tests, serotyping,
analysis of mycolic acid patterns by HPLC[2], and, if deemed necessary
(as was the case for 6 strains), nucleic acid sequence similarities
(DNA-DNA and RNA-RNA).

   When all tests were completed and it was confirmed that retesting
of cultures previously labeled "correct" did not alter their
identification, the data were reanalyzed (Fig. 5 and 6). One of the
cultures was non-acid-fast, and was deleted from the set, leaving 116
strains for comparison. A variety of reasons contributed to the initial
disagreements: a pigmented strain originally labeled M. scrofulaceum was
a pigmented M. intracellulare; 5 other strains originally placed in the
M. avium complex were shown to be other, taxonomically closely related
species; 2 initially reacted weakly with the probes because the latter
were used on the expiration date; 4 were misidentified because of faulty
interpretation of serotyping results; 1 strain was even shown to be a
mixture of both M. avium and M. intracellulare. In the final analysis,
35 M. avium cultures were specifically identified (Fig. 5), yielding
both sensitivity and specificity of 100%. Twenty strains were correctly
identified as M. intracellulare, and, again, both sensitivity and
specificity of the tests were 100%.


```
                         Final Identification
                            as M. avium

                                 +    -

        Gen-Probe       +      35    0

        Test Results    -       0   81

        Sensitivity:  35/35 = 100%
        Specificity:  81/81 = 100%
```

Figure 5. Final evaluation of M. avium-specific probe
           (Gen-Probe) against 55 confirmed M. avium
           complex strains (35 M. avium, 20 M.
           intracellulare) and 61 other mycobacteria.
           (Note: 1 non acid-fast culture omitted;
           see text)

Final Identification as
M. intracellulare

```
                          +    -
Gen-Probe        +       20    0
Test Results     -        0   96
```

Sensitivity: 20/20 = 100%
Specificity: 96/96 = 100%

Figure 6. Final evaluation of M. intracellulare-
specific probe (Gen-Probe) against 55
confirmed M. avium complex strains
(20 M. intracellulare, 35 M. avium) and 61
other mycobacteria. (Note: 1 non acid-fast
culture omitted; see text)

DISCUSSION

In our experience, all 4 of the Gen-Probe mycobacterial culture
confirmation kits provided extremely accurate identification of the
respective species (or genus) for which they were prepared. Routine use
of the 3 species-specific probes on more than 100 reference cultures
examined since the initial evaluation here reported has yielded 100%
agreement with both conventional and new research-identification
technology.

The most time-consuming step for probe culture-confirmation
is the preparation of individual cell suspensions. Although 2 or 3
unknown cultures can be examined in less than 3 hours, when as many as
20 cultures are tested in 1 day, an entire 8-hour work period may be
needed. One might argue that routine biochemical tests (niacin, nitrate
reduction, and 68% catalase) could identify most pure cultures of M.
tuberculosis within the same 2-3 hours needed for probe identification
of the same isolates, and this would be true. Perhaps the greatest uses
of the TB-probe have been in identifying variant cultures lacking 1 or
another of the 3 key biochemical features of the species and in
detecting M. tuberculosis on the same medium with other acid-fast or non
acid-fast (that is, mixed or contaminated) cultures. Also, the probe
may provide more reliable identification of M. tuberculosis for those
laboratories staffed by rotating personnel or by new microbiologists who
lack the experience to recognize the variant strains or the mixed
cultures.

In the case of the M. avium complex, the 2 new species
probes enable us now to identify specifically both M. avium and M.
intracellulare, where conventional methods formerly lumped the 2
together as M. avium complex. This will enable more precise assessment
of which species (if either) is more commonly associated with human
disease. Then, too, specific identification of the pure cultures by
probes can be realized within several hours, whereas conventional
methods take up to 4 weeks.

In summary, our experience in the evaluation of 4
commercially available DNA probes for culture confirmation of
mycobacteria (that is, probes to identify the genus Mycobacterium and
the species M. tuberculosis complex, M. avium, and M. intracellulare)
has been extremely gratifying, and continued day-to-day use in our
reference laboratory has only strengthened this feeling. One can but

hope that the probes currently projected to identify organisms directly
in sputum will perform as well as the culture confirmation probes; if
they do, precise identification of M. tuberculosis could be done within
the same day the sputum is collected. That is progress!

LITERATURE CITED

1. Butler, W.R., D.G. Ahearn, and J.O. Kilburn. 1986. High-performance
   liquid chromatography of mycolic acids as a tool in the
   identification of Corynebacterium, Nocardia, Rhodococcus, and
   Mycobacterium species. J. Clin. Microbiol. 23:182-185.
2. Butler, W.R. and J.O. Kilburn. 1988. Identification of major slowly
   growing pathogenic mycobacteria and Mycobacterium gordonae by high
   performance liquid chromatography of their mycolic acids. J. Clin.
   Microbiol. 26:50-53.
3. Butler, W.R., J.O. Kilburn, and G.P. Kubica. 1987. High-performance
   liquid chromatography analysis of mycolic acids as an aid in
   laboratory identification of Rhodococcus and Nocardia species. J.
   Clin. Microbiol. 25:2126-2131.
4. Edelstein, P.L. 1986. Evaluation of the Gen-Probe DNA probe for
   detection of Legionellae in culture. J. Clin. Microbiol. 23:481-484.
5. Good, R.C. and R.E. Beam. 1984. Chapter 5, Seroagglutination, pp
   105-122, in The Mycobacteria, A Sourcebook (Eds G.P. Kubica and L.G.
   Wayne), Marcel Dekker, Inc., New York.
6. Kent, P.T. and G.P. Kubica. 1985. Public health mycobacteriology: A
   guide for the Level III laboratory. HHS Publication No. (CDC)
   86-8230, Centers for Disease Control, Atlanta.
7. Kohne, D.E. 1986. Application of DNA probe tests to the diagnosis of
   infectious disease. American Clinical Products Review. November, 7
   pages.
8. Wilkinson, H.W., J.S. Sampson, B.B. Plikaytis. 1986. Evaluation of a
   commercial gene probe for identification of Legionella cultures. J.
   Clin. Microbiol. 23:217-220.

USE OF DNA PROBES FOR THE DIAGNOSIS OF INFECTIONS CAUSED BY MYCOPLASMA

PNEUMONIAE AND LEGIONELLAE- A REVIEW

Paul H. Edelstein

Departments of Pathology and Laboratory Medicine, and
Medicine
Hospital of the University of Pennsylvania
Philadelphia, PA

ABSTRACT

Specific DNA probes have been made for both M. pneumoniae and Legionella
species. Dot blot methods have been used in research laboratories to test
culture isolates of both organisms, and also to test animal tissues with a
L. pneumophila-specific probe. Commercial kits are also available for
direct specimen testing for these two organisms. The commercial kits are
made by a single manufacturer, Gen-Probe, Inc. (San Diego, CA), and use a
novel in-solution rapid hybridization assay, using ^{125}I-labeled cDNA to
rRNAs of the organisms. The Gen-Probe M. pneumoniae probe appears to be
80% to 100% sensitive, and 97% to 100% specific, based on analysis of two
clinical studies using positive culture as the diagnostic criterion. The
Gen-Probe legionella probe appears to be 33% to 71% sensitive (mean 57%),
and 98.9% to 99.7% specific (mean 99.7%), based on analysis of four prospective clinical studies, using positive culture as the definition of disease,
with a total sample size of 3,243 patients, 49 of which were culture-positive. Both Gen-Probe direct tests appear to be clinically useful, although
the poor performance of the legionella test in one major university laboratory, and the expense of performing these tests, mandate that thorough
evaluations be carried out in each laboratory anticipating using the test.
Culture must always be performed for legionella whether or not the DNA
probe test is used. It is likely that the use of the M. pneumoniae kit
would greatly speed diagnosis, but whether this would alter medical practice
or result in lower morbidity and health care costs is unknown.

INTRODUCTION

Diagnosis of both <u>Mycoplasma pneumoniae</u> respiratory tract infections and legionella pneumonia can be difficult on clinical grounds alone. Laboratory diagnosis of these diseases may require several days to weeks, and is not always specific (2,9). The reference method for the diagnosis of both diseases is culture, which appears to be the most sensitive and specific method. Rapid and specific diagnosis of legionella pneumonia can be performed by immunofluorescent microscopy, urine antigen detection, or both types of tests; while both of these types of tests are very specific for <u>L. pneumophila</u>, they have drawbacks of low sensitivity in some laboratories (immunofluorescence) and limited reactivity with other serogroups and species (immunofluorescence and urine antigen testing) (2,7). There is a great need for a rapid, sensitive and specific diagnostic test for legionella pneumonia, as misdiagnosis can result in considerable morbidity and even death. It is likely that an accurate and rapid test for mycoplasma pneumonia would also be beneficial, although misdiagnosis of this disease does not usually have such dire consequences. The prevalence of legionella pneumonia in adults with pneumonia is probably in the 1% to 5% range, making it necessary that any laboratory diagnostic test for this disease be very specific (>99.5%), whereas in certain populations, such as children and college age adults, the prevalence of <u>M. pneumoniae</u> pneumonia in those with pneumonia may be considerably higher (10% to 60%) such that a test of lower specificity may be acceptable.

GEN-PROBE ASSAYS

Methodology

The only commercially available and practical DNA probe assays for <u>M. pneumoniae</u> and legionellae in clinical samples are made by the Gen-Probe Corporation (San Diego, CA). As Dr. Kohne has described at this symposium, the Gen-Probe tests utilize a unique methodology which enables tests to be performed using a single tube assay. The test requires about three hour's time from start-to-finish, with most of the time being for incubations and only about 25 to 30 minutes being hands-on time. The Gen-Probe DNA probes utilize ^{125}I-labeled cDNA, reactive with specific rRNA's of the target organism(s). Bacteria present in throat swabs (mycoplasma), or lower respiratory tract secretions, fluids, and tissues (legionella), are lysed using a lysing reagent; in the case of legionella, sonication with glass beads is needed for complete lysis. The DNA probe is added to the lysed bacterial contents, and allowed to react under high strigency conditions (72° C). Double stranded nucleic acids are separated from single stranded (unreacted) nucleic acids by binding to hydroxyapatite, which is then washed free of unreacted cDNA. The washed pellet in the reaction tube is then counted in a gamma counter. The ratio of radioactivity of the sample to that of a negative control sample is determined. Sample-to-negative ratios ≥ 3.0 are considered positive for the mycoplasma assay, and those samples with ratios less than 3.0 are considered negative. The breakpoint for legionella is 7.1 (vide infra).

Equipment needed to perform these assays includes a high speed micro-centrifuge (mycoplasma), a table top centrifuge (mycoplasma and legionella), a precisely controlled hot water bath (both assays), a jeweler's heated sonicator bath (legionella), a micropipette, a vortex mixer, and a gamma counter.

Cost

Cost is a major consideration when performing these rapid assays. The shelf life of the kits is about six to eight weeks, because of the short half-life of ^{125}I. So regardless of use, new kits have to be purchased frequently. Negative and positive control samples must be run in each assay, making it much cheaper to batch samples than to test single ones. The kits contain enough reagent and tubes to perform 20 tests, but not to test 20 patient samples because of the need to use reagent and tubes for negative and positive control samples. The cost of performing the controls makes it exceptionally expensive to test a small number of samples at a time (Table 1). Counterbalancing this is the minimum incremental cost of testing additional specimens, such that the cost per test declines dramatic-ally once more when two patient specimens are tested at any one time.

Table 1. Estimated Costs to Perform
Gen-Probe Clinical Assays

No. Tests Performed	Costs Per Test[a]		
	Material[b]	Labor	Total
1	$ 38.00	5.00	43.00
2	25.00	3.00	28.00
3	21.00	2.00	23.00
5	18.00	1.50	19.50
7	16.00	1.50	17.50
10	15.00	1.50	16.50
18	14.00	1.25	15.25

[a]Inclusive of costs of testing control samples, but
exclusive of equipment costs.
[b]Calculated on the basis of the list price of $252.00
for a 20 test kit.

Table 2. DNA Probes for Mycoplasma

Assay Type	Label	Specificity	Sensitivity	Source	Reference
Dot blot Culture Confirmation	^{32}P, biotinylation, sulfonation	M. pneumoniae[a] M. genitalium[a]	10^5CFU	Research reagent only	5
Tube	^{3}H	Mycoplasma Acholeplasma Other prokaryotes?	??	Gen-Probe	NP[b]
Tube	^{125}I	M. pneumoniae and M. genitalium[c]	5×10^5CFU	Gen-Probe	NP[b]

[a]Separate probes for these two species, neither of which are cross-reactive.
[b]Not published.
[c]Probe binds highly to rRNAs of both species.

Equipment costs are not especially high, and are most dependent on whether
a single well ($2600) or multiwell ($25,000) gamma counter is used; considerable time and effort is saved when analyzing multiple specimens if a
multiwell counter is used. All necessary equipment, including a single well
gamma counter, can be purchased for about $7600, which is about the cost of
a microscope needed for immunofluorescent microscopy. Use of a Gen-Probe
DNA probe assay could be difficult to justify economically for laboratories
performing only 1 or 2 tests a day, whereas a laboratory testing in excess
of 7 to 10 specimens a day should be able to easily justify its use.

M. PNEUMONIAE

There is only one commercially available DNA probe test for detecting
M. pneumoniae in clinical samples (Gen-Probe) (Table 2). Gen-Probe also
makes a product designed to detect any mycoplasma or acholeplasma present
in tissue culture; this is of no utility clinically because it reacts with
saprophytic organisms. A dot blot assay specific for M. pneumoniae isolated
in culture has also been described; this uses ^{32}P, avidin-biotin, or sulfonated probes (6). The dot blot assay has not been tested with clinical samples, and is too cumbersome to use in a clinical laboratory.

No published data is available on the performance of the Gen-Probe DNA
probe for M. pneumoniae in clinical samples. Results of ongoing clinical
trials are shown in Table 3, which also shows data given by Gen-Probe in
the product brochure. The Gen-Probe data shows excellent, almost unbeliev-
able results, with 100% specificity and sensitivity. The method(s) of
confirming the diagnosis of M. pneumoniae infection is not stated, nor is
the method of specimen selection. Patients studied at the University of
Connecticut were of college age, and were suspected of having mycoplasma
pneumonia on clinical grounds. Positive culture for M. pneumoniae using
Hayflick's medium was used as the definition of disease; red cell hemolysis
was used as a confirmatory test. University students with acute upper and
lower respiratory tract illnesses were the subjects of the study at the
University of Pennsylvania. Growth of M. pneumoniac on SP-4 medium was
used as the definition of disease; identity of isolates was confirmed by
immunofluorescence. Reasons for differences in the sensitivity of the Gen-
Probe assay in these three studies is unclear, but might include differences
in disease duration before culture and presence or absence of pneumonia.
At least for the University of Pennsylvania population, the Gen-Probe assay
cannot replace culture, but there is a promise for that in other, perhaps
more highly selected, populations.

Because of the known chronic colonization state with M. pneumoniae
after infection, it may be diffcult to differentiate those with mycoplasma

Table 3. Clinical Performance of Gen-Probe M. pneumoniae Assay

Study Center	No. Patients	Sensitivity (%)	Specificity (%)	Culture Method
U. Connecticut[a]	84	100	98.4	Hayflick's Medium
U. Pennsylvania[b]	508	83.3	96.6	SP-4
Unknown[c]	388	100	100	Unknown[c]

[a]Unpublished data of R. Tilton and colleagues.
[b]Unpublished data of J. Kontra, I. Nachamkin, D. Smith, R. MacGregor, P. Edelstein.
[c]Unpublished data provided by Gen-Probe in the product brochure.

colonization from those with acute mycoplasma infections. Careful serial sampling of acutely ill patients will be needed to determine if the probe assay remains positive long after disease onset, or after initiation or termination of therapy.

Mycoplasma genitalium hybridizes very strongly to the Gen-Probe M. pneumoniae probe. This is unlikely to cause significant diagnostic confusion as the former species is exclusively a genital tract colonizer. However there is a possibility that sexually active patients may have transient oropharyngeal colonization with this organism, and could therefore have falsely-positive probe assays. This issue has not been addressed in any study.

Antimicrobial therapy appears to decrease Gen-Probe test sensitivity in the case of tuberculous infections (R. Enns, personal communication). It is unknown whether appropriate antimicrobial therapy for M. pneumoniae infections will have a similar effect. If so, a major effort will have to be made to collect specimens prior to antimicrobial therapy.

Will rapid testing for M. pneumoniae infections lower patient morbidity or costs of medical care? The answer is likely yes, but this has to be determined. For example, empiric erythromycin therapy may be much less expensive than the cost of laboratory testing for M. pneumoniae in certain high prevalence populations. Since M. pneumoniae infections tend only to be diagnosed specifically when patients have severe or unusal symptoms, it is unknown whether early antimicrobial therapy will ameliorate mild symptoms. thus, while it is evident that the Gen-Probe mycoplasma probe is remarkably sensitive and specific, it is unknown whether this testing will benefit all patients with M. pneumoniae infection.

LEGIONELLA

Several DNA probes have been made for legionellae. Two dot blot assays, using probes specific for L. pneumophila, have been described (4,5). Both assays utilized ^{32}P labeled DNA, and correctly distinguished L. pneumophila from other bacteria. One assay, developed by Engleberg and colleagues, was successfully used to detect L. pneumophila present in mouse lungs; this had a sensitivity of about 5×10^5 CFU (4). Neither of these assays would be practical for use in a clinical laboratory. Gen-Probe makes an assay kit for detection of legionellae in clinical specimens. Like the Gen-Probe assay for mycoplasma, this utilizes ^{125}I-labeled cDNA in a single tube assay, which takes about 3 hours to complete.

The Gen-Probe DNA probe for legionellae is genus specific. Some Legionella species hybridize more completely to the probe than do others,

Table 4. Results of Clinical Trials with the Gen-Probe Test
for Legionellae in Clinical Specimens

Study site	Study type	Number of Specimens	Number Cult. Positive	Break-point	Sensitivity (%)	Specificity (%)	Ref.
VA Wadsworth	Retrospective	342	112	4.0	57-74[a]	99.1-100[a]	3
UCLA	Prospective	210	3	7.1	67	100	b
Ohio State	Prospective	1377	15	7.1	33	99.9	c
Univ. Pitt.	Prospective	654	24	7.1	71	99.4	d
Harper Hosp.	Prospective	1002	7	7.1	57	99.7	e
All Sites	Prospective	3242	49	7.1	56	99.7	

[a]Lower figure is for all samples, and higher figure excludes samples which could not be recultured.
[b]Personal communication, D. Bruckner and T. Drake.
[c]Personal communication, W.J. Buesching.
[d]Personal communication, A.W. Pasculle.
[e]W.A. Paultke, B.B. Wentworth, J.G. Geiger, and G.D. Bostic. Abst. Ann. Meet. Am. Soc. Microbiol. 1987, No. 183.

with L. pneumophila hybridizing most completely, and L. bozemanii, L. oakridgensis, and L. wadsworthii least completely (1,10). The degree of hybridization of L. pneumophila is about three to four times that of the least reactive Legionella species. The probe can be used with 100% sensitivity and specificity to identify putative Legionella species isolates to the genus level, but cannot be used to differentiate between species.

There is only one published study of the utility of the Gen-Probe DNA probe for detection of legionellae in clinical specimens (Table 4) (3). This was a retrospective analysis using frozen specimens known to be culture-positive or culture-negative for legionellae. A major flaw of the study was the very limited amount of individual specimens available, which meant that simultaneous culture and immunofluorescence could not be performed. Overall, 57% of 112 culture-positive specimens had sample to negative ratios of >4.0, and 99.1% of 230 culture-negative samples had ratios less than 4.0. The false-negative probe samples were recultured if possible, and discarded from analysis if they were culture-negative or if there was no sample left for analysis. This changed the sensitivity to 74%; neither false-positive probe samples (both with ratios between 4 and 7) could be recultured to exclude legionella, and when these two samples were excluded from analysis, the specificity was 100%. Pleural fluids, transtracheal aspirates, and extrapulmonary organ specimens yielded false-negative probe assays more frequently than did other tissues and fluids. The ability of the probe to detect species of Legionella other than L. pneumophila could not be validly determined because of the relatively low numbers of specimens containing these other species, but at least some were positive.

Subsequent to completion of the retrospective study and production of non-research lots of probe kits, it became evident that the breakpoint of 4.0 did not provide the same degree of specificity as was observed in the original study. A pseudoepidemic of Legionnaires' disease in Alabama was found due to false-positive probe tests, all but one with ratios of 4 to 7 (S. Laussucq, D. Schuster, W.J. Alexander, H. Wilkinson, and J. Spika. Abst. Ann. Meet. Am. Soc. Microbiol. 1987, Abstract 182). Gen-Probe has now revised its guidelines and suggests that only specimens with sample-to-negative ratios >7.0 be called positive, those with ratios between 4 and 7 "equivocal", and those with ratios below 4 be called negative. If a ratio of >7.0 had been used in the retrospective study, the test sensitivity would have been 44% rather than 57%; calculation of sensitivity after sample exclusion is impossible because previously culture-positive samples with ratios between 4 and 7 were not recultured.

Table 5. Predictive Values of Gen-Probe DNA Probe Test for Legionellae in Clinical Specimens in Four Prospective Clinical Studies

Study Site[a]	Pneumonia Caused by Legionellae (%)[b]	Positive Predicitve Vaule (%)[c]	Negative Predicitve Vaule (%)[c]
UCLA	1.4	100	99.5
Ohio State Univ.	1.1	83	99.6
Univ. of Pittsburgh	3.7	81	98.9
Harper Hospital	0.7	57	99.7
Pooled Data	1.5	77	99.3

[a]See Table 4 for sources of data.
[b]As defined by positive culture.
[c]Using sample-to-negative ratio of >7.0 as definition of positive result, and ≤7.0 as negative.

Several prospective clinical trials of the Gen-Probe assay for legion-ellae in clinical specimens have now been completed or are still ongoing. These are summarized in Tables 4 and 5, using sample-to-negative ratios of greater than 7.0 as positive, and less than 7.0 as negative. In each case, a positive culture for legionella was used as the definition of disease. For all four prospective studies there was a total of 102 samples with ratios between 4.0 and 7.0; of these, ten (9.8%) were culture-positive and the remainder culture-negative. Thus odds are high that a sample with a ratio between 4 and 7 is truly-negative. The use of an "equivocal" zone should probably be dropped, and all samples with ratios less than 7.1 be called negative.

Comparison of results obtained at each study center is difficult, as each studied different types of samples and patients, and because different culture techniques were used at each hospital. UCLA and Presbyterian Hospital (University of Pittsburgh) used more than one selective plate and pre-plating acidification of specimens. Harper Hospital used two selective plates without pre-plating acidification, and Ohio State University Hospital used a single selective plate. Only Presbyterian and Harper Hospitals per-formed concurrent direct immunofluorescent testing. Harper Hospital tested sputum samples submitted for routine culture, whereas the other hospitals tested specimens obtained from patients with suspected Legionnaires' disease. Many of the samples tested as OSU were fluids or bronchial washings, which may have lower yield than sputum in the probe assay. Regardless of the study site, it is evident that the probe test cannot be substituted for cul-ture, but is specific enough to be used for screening for low prevalence disease, provided that breakpoint of 7.1 is used.

If rapid testing is to be offered by a laboratory, should the probe test or another rapid test, such as DFA or urine antigen testing, be used? Unfortunately, extensive data answering this question is not available. The DFA test was 99.9% specific and 69% sensitive as judged by culture results at Presbyterian Hospital, as opposed to 99.4% specificity and 71% sensitivity for the probe. At Harper Hospital, DFA testing was only 29% sensitive and 91% specific, as compared to 57% sensitivity and 99.7% speci-ficity for the probe; the DFA reagents used at Harper Hospital included many of unknown specificity and sensitivity. Based on historical results, probe sensitivity and specificity is about equal to those for DFA and urine antigen testing. The decision about which of these tests to use then becomes one of economics, technical expertise, equipment availability, work load, and commercial availability (for the urine antigen test). In a high volume laboratory testing many specimens daily for legionella, either the probe test or urine antigen testing would make the most sense. In a low volume laboratory, the availability of highly skilled immunofluorescent microscopists will be a major deciding factor.

An additional factor which needs to be added to the decision making process will be the relative frequency of disease caused by different Legionella species and serogroups, which varies geographically. The probe test is positive with at least some specimens containing legionellae other than L. pneumophila, and seems to be very good at detecting all serogroups of L. pneumophila. The same cannot be said for urine antigen testing, which is presently specific just for L. pneumophila serogroup 1. DFA testing using monoclonal antigen is specific for all L. pneumophila serogroups, but does not detect other species. DFA reagents specific for other Legionella species can be purchased, but their cross-reactivity with non-legionellae is probably greater than that observed with the monoclonal reagent. L. pneumophila probably causes about 80% of all legionella infections, and L. pneumophila serogroup 1 seems to cause the majority (80% to 90%) of L. pneumophila infections (8). Using a species or serogroup-specific reagent is a compromise which can be justified on economic grounds, and on the unknown ability of the DNA probe to detect other species in clinical samples.

CONCLUSIONS

Good quality DNA probes are now available for the direct detection of M. pneumoniae, L. pneumophila, and possibly other Legionella species. Neither of the commercial probes is as sensitive as culture, which must still be performed for optimal sensitivity. The sensitivity of the M. pneumoniae probe appears to be greater than that of the legionella probe. Several alternative rapid, immunologically-based, tests are available for legionella, which must be compared to the probe in standardized prospective studies. Economic issues may be more important than other factors in deciding whether or not to use the probe tests.

LITERATURE CITED

1. P. H. Edelstein, Evaluation of the Gen-Probe DNA probe for the detection of legionellae in culture, J. Clin. Microbiol. 23:481-484 (1986).
2. P. H. Edelstein, Laboratory diagnosis of infection caused by legionellae, Eur. J. Clin. Microbiol. 6:4-10 (1987).
3. P. H. Edelstein, R. N. Bryan, R. K. Enns, D. E. Kohne, and D. L. Kacian, Retrospective study of Gen-Probe Rapid Diagnostic System for detection of legionellae in frozen clinical repiratory tract samples, J. Clin. Microbiol. 25:1022-1026 (1987).
4. N. C. Engleberg, C. Carter, P. Demarsh, D. J. Drutz, and B.I. Einstein, A Legionella-specific DNA probe detects organisms in lung tissue homogenates from intransally inoculated mice, Isr. J. Med. Sci. 22:703-705 (1986).

5. P. A. Grimont, F. Grimont, N. Desplaces, and P. Tchen, DNA probe spec- ific for _Legionella pneumophila_, _J. Clin. Microbiol._ 21:431–437 (1985).

6. H. C. Hyman, D. Yogev, and S. Razin, DNA probes for detection and identification of _Mycoplasma pneumoniae_ and _Mycoplasma genitalium_, _J. Clin. Microbiol._ 25:726–728 (1987).

7. R. B. Kohler, Antigen detection for the rapid diagnosis of mycoplasma and _Legionella_ pneumonia, _Diag. Microbiol. Infect. Dis._ 4:47S–59S (1986).

8. A. L. Reingold, B. M. Thomason, B. J. Brake, L. Thacker, H. W. Wilkinson, and J. N. Kuritsky, Legionella pneumonia in the United States: the distribution of serogroups and species causing human illness, _J. Infect. Dis._ 149:819 (1984).

9. C. U. Tuazon, and H. W. Murray, Atypical Pneumonias, _in_: "Respiratory Infections: Diagnosis and Management," J. E. Pennington, ed., Raven Press, New York, NY, (1983), p. 251–267.

10. H. W. Wilkinson, J. S. Sampson, and B. B. Plikaytis, Evaluation of a commercial gene probe for identification of _Legionella_ cultures, _J. Clin. Microbiol._ 23:217–220 (1986).

MULTICENTER CLINICAL EVALUATION OF THE Du PONT HERPCHEK[TM] HSV ELISA, A NEW

RAPID DIAGNOSTIC TEST FOR THE DIRECT DETECTION OF HERPES SIMPLEX VIRUS

David A. Baker[1], Deborah Pavan-Langston[2],
Bernard Gonik[3], Peter O. Milch[1];
Edmund C. Dunkel[2], Albert Berkowitz[3],
Mary Beth Woodin[4], Abraham Philip[4],
Ronald J. Vander-Mallie[4], Mark N. Bobrow[4] and
George Cukor[4]

Department of Obstetrics and Gynecology, State University of
New York at Stony Brook, Stony Brook, N.Y. 11794,[1]
Eye Research Institute, Department of Ophthalmology, Harvard
Medical School, Boston, MA 02114,[2]
Department of Obstetrics and Gynecology, University of Texas
Health Science Center, Houston, TX 77030,[3]
Medical Products Department, E. I. du Pont de Nemours & Co.,
North Billerica, MA 01862,[4]

ABSTRACT

A new 4h rapid enzyme-immunoassay for direct detection of herpes
simplex virus (HSV) antigen (Du Pont HERPCHEK[TM]) was evaluated with 743
clinical samples collected at obstetrics and gynecology (OB/GYN), sexually
transmitted diseases (STD), and ophthalmology clinics. The sensitivity and
specificity of HERPCHEK[TM] was 98.0% and 98.4% respectively compared to
virus isolation in cell culture. Confirmatory blocking ELISA tests,
clinical history and follow up indicate that the true specificity of the
test is 100%.

INTRODUCTION

Diagnosis of herpes simplex virus infection is critical in several
areas of clinical medicine including obstetrics, genital infectious
diseases and ophthalmology.

The frequency of genital herpes simplex virus infection continues to
increase causing morbidity among adults and placing newborns at risk for
serious illness or death (2,9). HSV infection is responsible for about half
of the genital ulcerations among patients attending a gynecology or STD
clinic in the United States. Clinical diagnosis of these ulceration may be
made more difficult by the varied size, symptoms and appearance of HSV
lesions. A variety of genital ulcerations due to other infectious as well
as non-infectious causes must be differentiated from HSV infection (2). In
particular, a laboratory diagnosis is highly desirable in clinical
situations relating to neonatal HSV or where antiviral chemotherapy is
being considered.

In addition, HSV is also the leading cause of infectious uniocular blindness in the United States, with approximately 500,000 cases being reported annually (10,11). Typical cases of ocular HSV are readily recognized by slit lamp examination and observation of lesions. However, atypical cases such as conjunctivitis and keratoconjunctivitis lacking ocular lesions, and pediatric ocular HSV infections may complicate the differential diagnosis of HSV infection based solely on clinical observations. In light of the potentially serious complications of HSV ocular infection, and the availability of effective antiviral therapy, access to rapid and sensitive laboratory diagnosis is essential (14) and may alter the management and prognosis of the disease, particularly when corticosteroid therapy is an option.

HSV isolation in cell culture remains the "gold standard" for viral detection (3,7). However, some shortcomings of culture in managing the risk of neonatal herpes have now been recognized (1,5). In obstetrics as well as in a gynecology, STD, or ophthalmology setting, the usefulness of culture for HSV diagnosis is limited by such factors as availability, turn around time, transportation of sample to the laboratory and cost. Currently, commercially available rapid diagnostic tests play a relatively minor role in HSV diagnosis primarily due to their lack of sensitivity and consequent need for backing up negative results with culture (4,6,8,12,15,16).

This study presents the results of a multicenter clinical evaluation of a new highly sensitive herpes rapid diagnostic assay, Du Pont's HERPCHEK[TM] HSV ELISA test.

MATERIALS AND METHODS

A total of 563 genital and dermal samples were obtained in New York and Texas at two large OB/GYN clinics located in state university affiliated teaching hospitals and at an STD clinic in a county hospital. Patients were sampled if they had active lesions or a history of recurrent HSV infection. In addition, 180 ocular tear film samples were obtained at the cornea clinic of a medical school affiliated eye infirmary in Massachusetts. At the time of sampling, patients were examined and appropriate clinical and epidemiologic information was obtained.

Clinical samples were obtained using the HERPTRAN[TM] collection system, which is part of the HERPCHEK[TM] test kit. The HERPTRAN[TM] package consists of two sterile cotton tipped swabs and a tube of transport buffer. Specimens were collected by holding both swabs together in one hand and insuring that both swabs pick up adequate amounts of clinical material. One swab was placed into the HERPTRAN[TM] tube and saved for analysis by ELISA. A sample contained in HERPTRAN[TM] can be transported at room temperature. The other swab was placed into the institution's usual viral transport medium and sent to the virology laboratory for routine culture. For the collection of tear film specimens the lower and upper conjunctival cul-de- sac was swabbed. Swabs were retained in the fornix for 15 - 20 seconds to ensure maximum adsorption.

For viral culture 0.2 ml of specimen in VTM was inoculated into either confluent Vero or Primary Rabbit Kidney cell cultures. After adsorption, the monolayers were washed and refed with fresh medium consisting of Dulbecco Modified Eagle's Medium with 2% calf serum and antibiotics. Cultures were incubated for one week at 37°C in an atmosphere of 5% CO_2 and observed daily for the presence of cytopathic effect (CPE). When VTM specimens could not be cultured immediately after receipt by the virology laboratory, they were stored at -70°C (3,7).

The HERPCHEK^{TM} HSV ELISA (Du Pont, No. Billerica, MA) test was used according to the manufacturer's directions to test patient specimens. Each test run was validated with appropriate positive and negative controls. In principle, HERPCHEK^{TM} is a microtiter plate based forward sandwich ELISA. HSV antigen, contained in HERPTRAN^{TM} buffer, is captured on polystyrene wells coated with rabbit anti-HSV serum. Antigen is detected with an HSV-specific biotinylated monoclonal antibody reagent. The assay is developed by the sequential addition of streptavidin-horseradish peroxidase and o-phenylenediamine substrate. The results are read in a spectrophotometric plate reader at 490nm. Total assay time is approximately 4h. The test does not distinguish between HSV-1 and HSV-2.

Discrepant genital samples which were HERPCHEK^{TM} positive but tissue culture negative were retested by a confirmatory blocking ELISA. In this test, 100 ul of sample was preincubated for 30 minutes with 10 ul of anti-HSV serum. A second 100 ul portion of the sample was preincubated with serum lacking HSV antibodies. Following preincubation, the HERPCHEK^{TM} reactivity of the samples was determined. The reduction of O.D. signal by HSV antibody, compared with normal serum was expressed as percent blocking.

RESULTS

The study population at the OB/GYN and STD clinics consisted of 91% females and 75% whites with a mean age of 28 $\pm$ 8.5 years. A total of 563 samples were obtained with 83.5% of the samples being collected from the genital tract. The prevalence of HSV in the study population, as determined by culture isolation, was 12.8%.

Table 1 summarizes results of the evaluation of the HERPCHEK^{TM} test. Part A compares ELISA detection of HSV in clinical samples with HSV isolation in culture. For the OB/GYN and STD samples, good concordance with culture was found in that 484 samples were negative by both tests and 70 of the 72 culture positive samples were also ELISA positive. An additional 7 culture negative samples were found to be ELISA positive. The presence of HSV antigen in these 7 specimens was demonstrated by the confirmatory blocking HSV ELISA test. Each of the 7 samples displayed >90% blocking with specific anti-HSV serum. Three of the 7 samples were from crusted lesions where recovery of infectious virus is known to be least efficient (8,12,13). One of these patients had also received acyclovir therapy prior to culture.

Part B of Table 1, shows HERPCHEK^{TM} to have a sensitivity of 97.2% and 100% specificity for the genital and dermal samples. It calculates the performance of the ELISA based on samples which were demonstrated to have either infectious virus by culture or HSV antigen by a confirmatory blocking test. The specificity of the ELISA increased to 100% when more than one test was used to determine true positivity.

For the ocular study, samples were divided into four groups based on initial clinical diagnosis. Group 1 consisted of 41 samples from patients with active ocular herpetic infections. Of these, 30 presented clinically as HSV infection and 11 as acute or chronic varicella zoster (VZV) ocular infection. The second group consisted of 90 samples from patients with a history of ocular HSV but who were clinically normal at the time of sampling. Group 3 included 30 samples from patients with nonherpetic infections and group 4 was made up of 19 samples collected from healthy eyes. HERPCHEK^{TM} results completely agreed with a specialist's (DPL) clinical diagnosis. Of the 30 patients with clinically diagnosed acute HSV infection, all were HERPCHEK^{TM} positive and 27 were also culture positive

Table 1.   EVALUATION OF HSV DETECTION BY HERPCHEK^{TM}

A.   Comparison to cell culture:

	OB/GYN and STD	OCULAR	TOTAL
SENSITIVITY:	97.2% (70/72)	100% (27/27)	98.0% (97/99)
SPECIFICITY:	98.6% (484/491)	98.0% (150/153)	98.4% (634/644)
POS. PREDICTIVE VALUE:	90.9% (70/77)	90.0% (27/30)	90.7% (97/107)
NEG. PREDICTIVE VALUE:	99.6% (484/486)	100% (150/150)	99.7% (634/636)

B.   Comparison to cell culture, confirmatory ELISA and clinical history:

	OB/GYN and STD	OCULAR	TOTAL
SENSITIVITY:	97.2% (70/72)	100% (30/30)	98.0% (100/102)
SPECIFICITY:	100% (484/484)	100% (150/150)	100% (634/644)
POS. PREDICTIVE VALUE:	100% (77/77)	100% (30/30)	100% (107/107)
NEG. PREDICTIVE VALUE:	99.6% (484/486)	100% (150/150)	99.7% (634/636)

for HSV. Both HERPCHEK^{TM} and culture were negative for all the remaining samples. The three patients with positive ELISA and negative culture results were all in the late acute stage of infection (>7 days), and one of these individuals had started topical trifluoridine therapy. In all three cases the clinical evidence is clear that HSV infection was present. It is recognized that culture may be less than optimal for older or treated lesions. Again, as Table 1 indicates, the true specificity of HERPCHEK^{TM} appears to be 100%. In this study sensitivity was also 100%.

When both the ocular and genital/dermal studies are combined, the sensitivity and specificity of HERPCHEK^{TM} is found to be 98.0% (97/99) and 98.4% (634/644) respectively compared to virus isolation in cell culture. However, confirmatory tests and clinical history indicate that the actual specificity of the test is 100%.

DISCUSSION

In contrast to the traditional tissue culture method which recognizes infectious virus, HERPCHEK^{TM} detects viral antigens. Since it takes from 1 to 7 days to obtain culture results, culture provides information as to the presence of infectious virus in the patient only at the time of sample collection but not at the time the test results become available, since during the culture period the lesion has progressed or regressed. Prolonged antigen secretion in the absence of infectious virus has not been reported for HSV.

The data presented in this report indicates, that HERPCHEK^{TM} is similar in sensitivity to cell culture. Even though culture is not perfect, it is the most sensitive available laboratory test for HSV. The most realistic determination of specificity for HERPCHEK^{TM} utilizes, in addition to culture, a confirmatory blocking ELISA and clinical history to define true positivity. By these criteria, the assay is completely specific for HSV. Based on the results reported here, it appears that the use of HERPCHEK^{TM} will present no disadvantage over the use of culture in most clinical situations and provides the considerable advantages of timely and objective results and no sample degradation during transport.

LITERATURE CITED

1. Arvin, AM, PA Hensleigh, CG Prober, DS Au, LL Yasukawa, AE Witek, PE Palumbo, SG Paryani and AS Yeager. 1986. Failure of antepartum maternal cultures to predict the infant's risk of exposure to herpes simplex virus at delivery. N. Engl. J. Med. 315:796-800.

2. Corey, L. 1984. "Genital herpes" pp 449-474 In Sexually Transmitted Diseases, Holmes, KK, P-A Mardh, PF Sparling, PJ Wiesner (eds). McGraw-Hill, N.Y.

3. Drew, WL and WE Rawls. 1985. "Herpes simplex viruses" pp 705-710 In Manual of Clinical Microbiology (4th ed.) Lennette, EH, A Balows, WJ Hausler, HJ Shadomy (eds). American Society for Microbiology, Washington, D.C.

4. Goldstein, LC, L Corey, JK McDougall, E Tolentino, and RC Nowinski. 1983. Monoclonal antibodies to herpes simplex viruses: use in antigenic typing and rapid diagnosis. J. Infectious Diseases 147:829-837.

5. Gordon, WA, L Apodaca, L Cragun, EM Peterson and LM de la Maza. 1987. Neonatal herpes simplex virus infection occurring in second twin of an asymptomatic mother: failure of a modern protocol. JAMA 257:508-511.

6. Halstead, DC, DG Beckwith, RL Sautter, L Plosila and KA Schneck. 1987. Evaluation of a rapid latex slide agglutination test for herpes simplex virus as a specimen screen and culture identification method. J. Clin. Microbiol. 25:936-937.

7. Hsiung, GD 1982. Diagnostic Virology. Yale University Press, New Heaven.

8. Lafferty, WE, S Krofft, M Remington, R Giddings, C Winter, A Cent and L Corey. 1987. Diagnosis of herpes simplex virus by direct immunofluorescence and viral isolation from samples of external genital lesions in a high-prevalence population. J. Clin. Microbiol. 25:323-326.

9. Morbidity and Mortality Weekly Report. 1986. Centers for Disease Control, Atlanta. Ann. Summary 33:90.

10. Nesburn, AB, and MT Green. 1976. Recurrence in ocular herpes simplex infection. Invest. Ophthalmol. Vis. Sci. 15, 515-518.

11. Pavan-Langston, D. 1979. Current trends in therapy of ocular herpes simplex: Experimental and clinical studies. Adv. Ophthalmol. 38:82-88.

12. AR Solomon, JE Rasmussen, J Varani and CL Pierson. 1984. The Tzanck smear in the diagnosis of cutaneous herpes simplex. JAMA 251:633-635.

13. Spruance, SL, JC Overall, ER Kern, GG Krueger, V Pliam and W Miller. 1977. The natural history of recurrent herpes simplex labialis. N. Engl. J. Med. 297:69-75.

14. Walpita, P, S Darougar, and U Thaker, 1985. A rapid and sensitive culture test for detecting herpes simplex virus from the eye. Br. J. Ophthalmol 69:637-639.

15. Warford, AL, RA Levy, and KA Rekrut. 1984. Evaluation of a commercial enzyme-linked immunosorbent assay for detection of herpes simplex virus antigen. J. Clin. Microbiol. 20:490-493.

16. Warford, AL, RA Levy, KA Rekrut and E. Steinberg. 1984. Herpes simplex virus testing of an obstetric population with an antigen enzyme-linked immunosorbent assay. Am. J. Obstet. Gynecol. 154:21-8.

NON-CULTURE TESTS FOR THE DIAGNOSIS OF GONORRHEA

Leonard Zubrzycki

Department of Microbiology and Immunology
Temple University School of Medicine
Philadelphia, PA

INTRODUCTION

Gram-stain examination and culture of clinical specimens are the standard diagnostic tests for gonorrhea. However, both techniques have limitations. Gram stain is relatively insensitive for the diagnosis of cervical
and rectal infections, and correct interpretation of the Gram stain requires
an experienced reader.[1] Culture results may be unsatisfactory, unless selective culture medium is quality controlled, stored, and inoculated properly,
and promptly incubated in a CO_2-enriched atmosphere. Also the culture method
using the common selective medium fails to detect vancomycin-susceptible
strains of Neisseria gonorrhoeae. Further problems arise when specimens must
be sent to another facility for culturing.

To overcome the limitations of Gram-stain examination and culture procedures, other non-culture tests were developed. Results using those non-
culture tests will be reviewed.

PREDICTIVE VALUES OF A DIAGNOSTIC TEST

Vecchio[2] provided a formula for estimating the predictive value (PV) of
a diagnostic test result. The PV is based on the sensitivity and specificity
of the test, and the prevalence of the disease in the population being
studied. What often occurs in diagnostic tests are false positive results
which are magnified when the disease is low in prevalence. To minimize such
errors a diagnostic test should have a high specificity even at the expense
of a lower sensitivity. A high specificity results in a high predictive
value positive (PVP). Because gonorrhea is not a high prevalence disease
relative to the number of tests (cultures, Gram staining) performed each
year, the specificity and PVP of the non-culture tests being reviewed will be
presented where appropriate.

Culture results are used as the standard for determining the sensitivity
and specificity of a diagnostic test for gonorrhea. Because the culture
method is not 100% sensitive, the PVP of the non-culture tests being reviewed
are probably higher than presented.

SEROLOGIC TESTS

There have been many serologic tests, using a variety of antigens,
developed for the diagnosis of gonorrhea. Koransky and Jacobs,[3] and Donegan[4]
reviewed those tests and concluded that a reliable serologic test for

Rapid Methods in Clinical Microbiology
Edited by B. Kleger *et al.*
Plenum Press, New York

gonorrhea is not available. The tests are unreliable because: there are antibodies in non-infected individuals; antibodies persist after therapy for gonorrhea; the antigenic makeup of gonococcal strains varies; there are different antibody responses in men and women, and in different races.

As an example of the heterogeneity of antibody responses in different groups of people, Donegan presented the unpublished study by M. L. Zeckel, L. E. Warfel, J. H. Reynolds, and G. F. Brooks. The study was on 655 sera from 372 patients. The serologic test was a radioimmunoassay for antibody to a purified gonococcal pilus protein. Zeckel et al. studied the antibody level in: black and white races; males and females; those uninfected, infected for the first time, and repeatedly infected. The geometric mean antibody levels overlapped among the four groups of subjects, and between those having gonorrhea, for the first or repeated times, and those not having gonorrhea. The conclusion is that it would be difficult to set an antibody titer indicative of a positive test for gonorrhea.

TABLES

The values in the tables which follow are those in the publications cited or calculated by this reviewer from the data in those publications. That is the reason why some values are in whole numbers and others are presented to the first decimal point.

The abbreviations used are: No. Spec, number of specimens tested; Prev%, the percent of subjects who were culture positive for gonorrhea; Sens%Spec, the % sensitivity and % specificity of the test relative to culture results; PVP%, the predictive value positive expressed in percent; Ref, reference(s).

LIMULUS AMOEBOCYTE LYSATE ASSAY

The Limulus amoebocyte lysate assay (LALA) detects endotoxin in body fluids. A test kit is commercially available (Gonoscreen, Mallinckrodt, St. Louis, MO). A body fluid is added to the liquid lysate, which then is incubated at 37°C. If the body fluid contains a sufficient quantity of endotoxin a gel forms in about a half hour. A positive test can be seen in about 10 minutes if a chromogenic substrate is used. The LALA is rapid and intended to replace the Gram stain for those unskilled in preparing and reading the Gram-stained preparations. The results of LALA on urethral discharge compared to culture results are shown in Table 1. They are taken from the studies of Prior and Spagna,[5] and Judson et al.[6] The results look good. However, they were obtained with a select population of men with a copious discharge of about 15-25 µl. Such a volume is necessary to perform the test. This limited the test to about 70% of the subjects entered into the studies reported in Table 1.

Judson et al. pointed out that the Gram stain could be performed on all men in the study and not just 70%. Because of the lack of discharge, the LALA cannot be used on asymptomatic men, or for test-of-cure. The LALA cannot be used with specimens from any other anatomic site, because of the presence of endotoxins of the normal flora at those sites.

Table 1. Results With Gonoscreen on Male Urethral Discharges

No. Spec	Prev%	Sens%Spec	PVP%	Ref
200	40	95 96.7	95	5
206	81	99.5 95.8	99	6

Table 2.   Results With Gonodecten on Male Urethral Discharges

No. Spec	Prev%	Sens%Spec		PVP%	Ref
201	66	95	60	82.2	7
240	76	96	84	95.0	8

THE OXIDASE TEST FOR GONOCOCCAL URETHRITIS IN MALES

Another rapid test intended to substitute for the Gram stain on urethral dicharges from men is an oxidase test. Such a test is commercially available (Gonodecten, U. S. Packaging Co., La Porte, IN). It is a simple test. A swab with male urethral discharge is put into a plastic tube which contains dry oxidase reagent and an ampoule with saline. When the ampoule is crushed the saline wets the dry oxidase reagent. Varying degrees of purple color develop on the swab within minutes, indicating a positive test. Table 2 shows the results from two studies comparing Gonodecten results with those from culturing. Felman and William[7] found that the oxidase test was as reliable as the Gram stain (results not shown). Janda and Jackson[8] found that the Gram stain was superior in being 99.5% sensitive and 100% specific.

Gonodecten is subject to false positive results due to iron-containing compounds such as heme from lysed erythrocytes. Also false positive results could be due to other oxidase positive bacteria. Therefore, the test cannot be used on cervical, rectal or pharyngeal swabs. Also the test cannot be used for men with minimal or no discharge. As suggested by Janda and Jackson,[8] the greatest utility of the test would be as a rapid screen, with Gram staining and culturing being performed on negative or equivocal oxidase tests.

DETECTION OF A SPECIAL GONOCOCCAL ENZYME

Takeguchi et al.[9] developed a test to detect 1-2 propanediol oxidoreductase in clinical material. This enzyme is found in appreciable concentrations only in Neisseria and Acinetobacter. In the presence of NAD, the enzyme converts 1-2 propanediol to an unidentified product which can be detected by a fluorometer. Table 3 shows the results from their study. The investigators were encouraged because the results were at least as good as those reported for serologic tests.

Some of the false positive results with cervical specimens could be due to cervical tissue, unknown enzymes, methodology, and, they state, even false negative culture results. They do admit that further studies would be needed to perfect this method for the diagnosis of gonorrhea.

Table 3.   Detection of 1-2 Propanediol Oxidoreductase in Cervical and Male Urethral Specimens[a]

	No. Spec	Prev%	Sens%Spec		PVP%
male	331	37.8	81.6	95.2	91.2
female	217	20.7	77.8	68.7	39.4

[a]Takeguchi et al.[9]

RADIOIMMUNOASSAY FOR GONOCOCCAL ANTIGENS IN URINE

Thornley et al.[10] developed a radioimmunoassay (RIA) to detect gonococcal antigens in urine. The antibody was IgG from a rabbit injected with six
isolates of gonococci. The IgG was labeled with radioactive sodium iodide
and coupled to cellulose. Gonococcal antigens in urine bind to the
cellulose-antibody complex, which is washed, centrifuged, and counted in a
gamma counter.

The RIA was positive with 31 of 42 urines from men known to be culture
positive for gonorrhea (74% sensitivity). The same cutoff value between
positive and negative results established with urine from men was used with
urine from women who were known to be culture positive or negative for gonorrhea. RIA was positive with 10 of 14 urines (71% sensitivity) from culture
positive women, but also positive with 3 of 18 urines (83% specificity) from
culture negative women. The problems with the RIA method as described are:
the high background binding of IgG antibody to urine sediments even in uninfected subjects; the methodology is complex; the antigenic heterogeneity of
gonococci; the use of radioisotopes.

GAS LIQUID CHROMATOGRAPHY FOR GONOCOCCAL FATTY ACID

Another novel method for detecting gonococci in clinical material is to
assay for 3-hydroxy dodecanoic acid, a major component in the fatty acid content of Neisseria.[11] Although Neisseria other than N. gonorrhoeae have a
high content of this fatty acid, the investigators relied on the reports
that those Neisseria species are not present in large numbers in cervical
material. The investigators extracted the fatty acids of a specimen and then
used gas liquid chromatography (GLC) to get the fatty acid profile of the
extract. Using GLC analysis, they correctly identified 15 of 19 cervical
specimens from women known to be culture positive (79% sensitivity) and 12 of
15 cervical specimens from women known to be culture negative (75% specificity). The problems with this test are: the limit of sensitivity is about
10^5 colony forming units, and often fewer are found in cervical material;
cervical material interferes with the test, necessitating complex purification procedures.

DETECTION OF GONOCOCCAL ANTIGENS BY ELISA

Young et al.[12] developed an ELISA test for detecting gonococcal antigens
in clinical material. However, most of the studies using an ELISA approach
have been done with a commercial test, Gonozyme (Abbott Diagnostics, Chicago,
IL).

Table 4. Results With Gonozyme for Detecting Gonococcal
Antigens in Male Urethral Specimens

No. Spec	Prev%	Sens%Spec		PVP%	Ref
465	36	95	98	97	13
664	2	67	98	30	13
208	54	97	95.8	96.5	14
117	64	93	100	100	15
151	28	96	99.1	97.5	15
198	56	99	92.5	93.7	16
419	20	95	99.4	97.6	17

Table 5. Results With Gonozyme for Detecting Gonococcal
Antigens in Cervical Specimens

No. Spec	Prev%	Sens%	Spec	PVP%	Ref
723	15	78	98	85	13
252	29	79	87.2	71.3	14
281	35	66	97.8	94.3	15
171	14	87	91.2	61.6	15
119	28	89	89.5	69.7	17
158	22	100	70.7	49.3	17
1,792	8.3	87	89.1	42	18

Table 4 is a composite of typical results obtained with male urethral
specimens. Overall, the results look good. With regard to the results
reported by Stamm et al.[13] with a 2% prevalence of disease: even if the
sensitivity remained at 67% but the specificity were 99.5% or, better yet,
99.9%, the PVP would be 73% and 93%, respectively. This emphasizes the point
made at the beginning of this review. Use a test with a high specificity.

The results with cervical specimens, shown in Table 5, are not as good
as those with urethrals. The lower sensitivity is perhaps due in part to
less antigen in cervical material. An additional explanation was offered
by Papasian et al.[14] The bead of the Gonozyme test system used to adsorb
gonococcal antigen perhaps adsorbs nongonococcal antigens from the complex
microbial flora in some cervical specimens. This could interfere with the
attachment of the anti-gonococcal antibody. Conversely, the polyclonal anti-
gonococcal antibody used in the test might cross-react with those antigens,
accounting for the lower specificity of the Gonozyme test with specimens from
the female.

Gonozyme has been used to detect antigens in urine specimens of males.
Rudrik et al.[19] found that the sensitivity and specificity on first-voided
uncentrifuged urine were 91.6% and 97.9%, respectively. Schachter et al.[20]
found that the sensitivity and specificity on urine sediment were 93% and
99%. Both groups of investigators agree that Gonozyme testing of urine is
perhaps adequate for the diagnosis of gonorrhea in males.

Table 6. Comparison of Gonozyme and Culture
for Test-of-Cure Evaluation

Culture	Gonozyme	Ref 14[a]	Ref 17[b]	Ref 18[c]
+	+			
+	−			
−	+		10	29
−	−	39	37	87

[a]24 males, 15 females
[b]37 males, 10 females, the discordant results
[c]116 females

Table 6 shows the results of using Gonozyme for test-of-cure evaluations. In two of the three studies gonococcal antigens persisted longer than viable gonococci. Gonozyme appears not to be reliable for test-of-cure evaluations of females.

The technology of the ELISA test is familiar to a serology laboratory. Therefore, Gonozyme results could be obtained at the site of collection, the same day a specimen is taken. This means that treatment could begin the same day. Specimens can also be sent to a central laboratory for processing. This is a desirable feature where processing of specimens cannot be done in-house. It is known that the culture method is not 100% sensitive. Actually, in a few of the studies[13,16,17] the Gram-stain results with male urethral exudates suggest that several of the false positive Gonyzyme results could be due to false negative culture results. This means that the specificity and PVP of Gonozyme might be greater with male urethral specimens than reported in Table 4. Nevertheless, it would be desirable to have a test with a higher specificity and PVP, especially for cervical specimens.

DNA HYBRIDIZATION FOR GONOCOCCAL URETHRITIS IN MALES

Totten et al.[21] and Perine et al.[14] used a ^{32}P-labeled 2.6 megadalton (Md) cryptic plasmid of N. gonorrhoeae as a probe to detect gonococcal DNA in male urethral exudate. The results in Table 7 are those obtained after exposing the DNA hybrids, trapped on nitrocellulose filter paper, to x-ray film at $-70^{\circ}C$ for three days.

Perine et al.[22] also used a ^{32}P-labeled 4.4 Md beta-lactamase plasmid as a probe to detect similar plasmid DNA in the male urethral exudate. The results from the DNA hybridization method for beta-lactamase were compared to the standard chromogenic cephalosporin test on gonococcal isolates from the urethral exudates. The sensitivity and specificity of detecting a beta-lactamase plasmid in specimens compared to the results with isolates available from those specimens were 91% and 98%, respectively.

Perine et al.[22] summarized the advantages and limitations of DNA hybridization for gonococcal urethritis in men. The specimen can be kept on nitrocellulose filter paper and sent from remote distances to a central laboratory for processing. They state that the cost of DNA hybridization is lower per specimen versus that of culturing at their facility. It is not clear, however, whether the cost includes the labor of preparing probes every two weeks because of the short half-life of ^{32}P. Nor is it clear whether this includes the cost of the equipment and facilities for handling radioisotope probes and detecting the DNA hybrids. DNA hybridization does allow the detection of beta-lactamase in clinical material, which detection cannot be done by other non-culture methods.

Perine et al.[22] point out that not all gonococci contain the cryptic plasmid. The absence of such plasmid would result in a false negative test. The DNA hybridization test for the gonococcal cryptic plasmid is less sensi-

Table 7. Results With the DNA-Hybridization Method for the Diagnosis of Gonococcal Urethritis in Males

No. Spec	Prev%	Sens%Spec		PVP%	Ref
113	63	89	100	100	21
216	87	96	93	98.8	22

tive with cervical and rectal specimens. The beta-lactamase probe would
probably not be useful on specimens from the cervix, rectum or throat,
because of the presence of beta-lactamase producing Haemophilus influenzae
and Escherichia coli with plasmids with DNA homologous to that in N.
gonorrhoeae.

The author of this review would add another disadvantage to the DNA
hybridization tests as described. It is the use of ^{32}P in a diagnostic test.
This subjects the laboratorian to unnecessary exposure to radiation. A
biotin-labeled DNA probe with an avidin-enzyme detection system would be more
desirable for safety, and for stability of the reagent probe. However, such
a probe might be less sensitive than a radioisotope probe.

DETECTION OF GONOCOCCAL DNA BY A GENETIC TRANSFORMATION TEST

Janik et al.[23] were the first to demonstrate the feasibility of using
a genetic transformation test for detecting gonococcal DNA in clinical
material. Apparently, their type of test using an auxotrophic variant to
detect gonococcal DNA in clinical material lacks sensitivity and specificity.[12]

Gonostat (Technology Management and Marketing, Inc., Santa Clara, CA) is
a different genetic transformation test (GTT). For this transformation test
a growth deficient mutant of N. gonorrhoeae is used, ATCC 31953. Briefly, a
DNA extract of clinical material collected on a cotton swab is spotted onto a
lawn of the mutant on a chocolate agar plate. The plate is incubated at $37^{\circ}C$
in a CO_2-enriched atmosphere. The mutant does not grow well under those conditions. A positive test for gonorrhea depends on gonococcal DNA in the
extract of clinical material to transform the mutant so that it will grow
into visible colonies. A negative test is seen as the absence of colony
growth.

The results shown in the following tables were obtained with duplicate
specimens collected at the DeKalb County Clinic, Atlanta, Ga., and the
Peoria Health Department, Peoria, IL. One specimen was cultured in-house
and the other was sent by mail for GTT. No refrigeration or holding media
are needed. The average length of time between collecting a specimen and
performing the GTT was five days.

The results in Table 8 are those with male urethral specimens. The
sensitivity of the test is good. GTT even detected the seven culture positive asymptomatic men.[24] But more important is the specificity. The specificities in parentheses are adjusted based on Gram-stain results and other
evidence for gonorrhea which will be presented later in this review. In the
study by Whittington et al.,[25] specimens were cultured in triplicate or
quadruplicate on culture medium with vancomycin, and others without. In that
study GTT detected 101% of the positives compared to those detected on the
vancomycin-containing medium.

Table 8. Results With Gonostat for Detecting Gonococcal DNA
in Male Urethral Specimens

No. Spec	Prev%	Sens%Spec		PVP%	Ref
72	38.7	97	98.3 (99.9)	97.4	24
367	29.6	94	98.9 (99.7)	97.3	25

Table 9. Results With Gonostat for Detecting Gonococcal DNA
in Cervical Specimens

No. Spec	Prev%	Sens%Spec		PVP%	Ref
254	40.9	97	98.7 (100)	98	24
492	22.5	85	98.4 (99.2)	94	25

Table 9 shows the results with cervical specimens. The GTT is less sensitive with those specimens than with male urethral specimens. However, the specificities are about the same, especially when they are adjusted, as shown in parentheses, to take into consideration possible false-negative culture results. In the study by Whittington et al.,[25] the sensitivity of of GTT on cervical specimens relative to culturing on vancomycin-containing medium was 90%.

Table 10 shows the results with the anal (rectal) specimens from females. The sensitivities are lower but the specificities are about the same as those with male urethral specimens. The sensitivity of the GTT with anal specimens remained the same when compared to vancomycin-containing medium.

Table 10. Results With Gonostat for Detecting Gonococcal DNA
in Female Anal Specimens

No. Spec	Prev%	Sens%Spec	PVP%	Ref
160	16.9	96.3 97 (100)	86.7	24
255	8.2	85.7 97.9 (98.3)	78.5	25

Table 11. Composite Gonostat Results[a]

	No. Spec	Prev%	Sens%Spec		PVP%
Male urethral	746	33.9	96	98.6 (99.8)	97.3
Cervical	796	30.4	92	98 (99.5)	95.3
Female anal	415	11.6	92	97.5 (98.9)	82.8

[a]Results reported in Tables 8-10 plus unpublished results from two pilot studies.

In the study by Jaffe et al.[24] it was shown that deliberately moistening the swab after the specimen was taken reduced the sensitivity of the test. It was intended that there be no deliberate moistening of swabs before or after collecting a specimen. However, all the rectal swabs collected at the DeKalb County Clinic for the study by Whittington et al.[25] were moistened. Those results with moistened swabs were also excluded from the results shown. However, in both studies the unadjusted specificities for the results with moistened swabs were still greater than 98%.

The composite results of the two studies evaluating the GTT and two pilot studies are shown in Table 11. The pilot studies were also blind studies in that the culture and Gram-stain results obtained at the DeKalb clinic were not revealed until the GTT results were completed.

The following evidence indicates that some of the false positive GTT results were probably due to false negative culture results.

There were seven apparent false-positive GTT results with male urethral specimens: six had intracellular gram-negative diplococci. That leaves one unexplainable result with a specimen from a patient who had urethritis.

There were 11 apparent false-positive GTT results with cervical specimens: six came from patients who had sex partners with gonorrhea; one had gram-negative intracellular diplococci, and the patient had a sex partner with gonorrhea; one came from a patient who had a positive anal culture and a positive anal GTT. That leaves three unexplainable results.

There were nine apparent false positive GTT results with anal specimens from females: two came from patients who had positive cervical cultures and positive cervical GTT, and the patients had sex partners with gonorrhea; one came from a patient who had a sex partner with gonorrhea; two came from patients who were diagnosed at another clinic as having gonorrhea. That leaves four unexplainable results.

With a culture sensitivity from males of about 95% to 98%, and about 85% in females (references in Stamm et al.[13]), the small percentage of difference between the results of culture and Gonostat can never be resolved.

The information just presented is the justification for the adjusted specificities shown in Tables 8-11. Fortunately, in the Whittington et al.[25] study, multiple platings on culture media were done to detect vancomycin-sensitive strains, or else there would be 15 more apparent false positive GTT results which would need explanation, because the GTT detected those 15.

Table 12 shows the results of test-of-cure evaluations on 68 males in the study by Jaffe et al.,[24] and on 74 males and 86 females in the study by Whittington et al.[25] In the latter, cervical and anal specimens were taken from all but two patients.

There were 3 discordant results from two female patients. From one patient the GTT was negative with the cervical specimen but positive with the rectal specimen, whereas culture was positive with both specimens. That leaves only one patient of 228 from whom there were discordant results. From that patient the cervical and rectal specimens yielded 3 and 2 colony-forming units, respectively. Although the colonies were presumptively identified as gonococci, the isolates were lost upon subculture and thus not confirmed.

Gonostat is not intended for in-house use. Care must be taken to maintain the test strain as colony type 1 for maximum competence for transformation. The GTT cannot detect infection with penicillinase-producing strains or those with chromosomal mutations for antibiotic resistance. The GTT cannot differentiate Neisseria meningitidis from N. gonorrhoeae, and therefore cannot be used on pharyngeal specimens.

Table 12. Comparison of Gonostat and Culture
for Test-of-Cure Evaluation

Test results			
Culture	Gonostat	Ref 24	Ref 25
+	+		5
+	−		3
−	+		
−	−	68	236

However, as pointed out by Jaffe et al.[24] and Whittington et al.,[25] the GTT has a high specificity and PVP. Those authors conclude that when conventional culture is not available and specimens must be sent a distance, the GTT is an attractive alternate.

CONCLUSION

The tests for gonorrhea based on detecting endotoxin, enzymes, or a certain fatty acid are non-specific and not practical. The Limulus amoebocyte lysate assay, Gonoscreen, and the test for oxidase, Gonodecten, are limited to males, specifically those with a copious discharge.

A test result depending on an antigen-antibody reaction, such as a serologic test to detect antibody, and Gonozyme to detect antigen, will always be compromised by any adjustment to increase sensitivity, because it will be done at the expense of specificity, and vice versa. The potential usefulness of monoclonal antibodies to detect antigens of gonococci in clinical material is beyond the scope of this review, which is to summarize existing tests.

The in vitro DNA hybridization test for gonorrhea is limited to male urethral specimens. Perhaps clinical material interferes with the sensitivity of the test for cervical and anal specimens. The only advantage this test has over the other non-culture tests, including the Gram stain, is that it can detect beta-lactamase in clinical material. This advantage must be weighed against the disproportionate cost to test for the small percent of cases of gonorrhea in the U.S.A. due to beta-lactamase-producing strains. The surveillance program on a sample of gonococcal isolates seems to be a reasonable and cost effective approach to determine the antibiotic resistance profile of gonococci in a community.[26]

The genetic transformation test, Gonostat, is an in vivo DNA hybridization test resulting in general genetic recombination. It is a natural test system in that the "detector" for gonococcal DNA is a strain of gonococcus. That is probably the reason why mucus, blood, and exudate of clinical material are less likely to interfere with the "detector." The specificity of the test, even the sensitivity, both within the 95% confidence limits of culture results, make it an attractive alternate when specimens must be transported for the diagnosis of gonorrhea.

LITERATURE CITED

1. M. E. Goodhart, J. Ogden, A. A. Zaidi and S. T. Kraus, Factors affecting the performance of smear and culture tests for the detection of Neisseria gonorrhoeae, Sex. Trans. Dis. 9:63 (1982).

2. T. J. Vecchio, Predictive value of a single diagnostic test in unselected populations, New Eng. J. Med. 274:1171 (1966).

3. J. R. Koransky and N. F. Jacobs, Serologic testing for gonorrhea, Sex. Trans. Dis. 4:27 (1977).

4. E. A. Donegan, Serological tests to diagnose gonococcal infections, In: "Gonococcal Infection," G. F. Brooks and E. A. Donegan, eds., Edward Arnold, London (1985).

5. R. B. Prior and V. A. Spagna, Improved utility of gonoscreen, a Limulus amoebocyte lysate assay, in the evaluation of urethral discharges in men, J. Clin. Microbiol. 22:141 (1985).

6. F. N. Judson, B. A. Werness and M. R. Shahan, Lack of utility of a limulus amoebocyte lysate assay in the diagnosis of urethral discharges in men, J. Clin. Microbiol. 21:152 (1985).

7. Y. M. Felman and D. C. William, Gonodecten - new 3-minute in vitro diagnostic test for gonorrhea in the male without use of conventional culture or gram stain, Urology, Vol. XIX (1982).

8. W. M. Janda and T. Jackson, Evaluation of Gonodecten for the presumptive diagnosis of gonococcal urethritis in men, J. Clin. Microbiol. 21:143 (1985).

9. M. M. Takeguchi, H. H. Weetal, D. K. Smith, H. C. McDonald, K. A. Livsey, C. C. Detar and T. A. Chapel, Enzymatic detection of Neisseria gonorrhoeae, Br. J. Vener. Dis. 56:304 (1980).

10. M. J. Thornley, D. V. Wilson, R. Demarco De Hormaeche, J. K. Oates and R. R. A. Coombs, Detection of gonococcal antigens in urine by radioimmunologyassay, J. Med. Microbiol. 12:161 (1979).

11. I. J. Sud and D. S. Feingold, Detection of 3-hydroxy fatty acids at picogram levels in biological specimens. A chemical method for the detection of Neisseria gonorrhoeae?, J. Invest. Dermatol. 73:521 (1979).

12. H. Young, S. K. Sarafian, A. B. Harris and A. McMillan, A non-cultural detection of Neisseria gonorrhoeae in cervical and vaginal washings, J. Med. Microbiol. 16:183 (1983).

13. W. E. Stamm, B. Cole, C. Fennell, P. Bonin, A. S. Armstrong, J. E. Herrmann and K. K. Holmes, Antigen detection for the detection of gonorrhea, J. Clin. Microbiol. 19:399 (1984).

14. C. J. Papasian, W. R. Bartholomew and D. Amsterdam, Validity of an enzyme immunoassay for detection of Neisseria gonorrhoeae antigens, J. Clin. Microbiol. 19:347 (1984).

15. J. Schachter, W. M. McCormack, R. F. Smith, R. M. Parks, R. Bailey and A. C. Ohlin, Enzyme immunoassay for diagnosis of gonorrhea, J. Clin. Microbiol. 19:57 (1984).

16. M. A. Nasello, D. R. Callihan, M. A. Menegus and R. T. Steigibel, A solid-phase enzyme immunoassay (Gonozyme) test for direct detection of Neisseria gonorrhoeae antigen in urogenital specimens from patients at a sexually transmitted disease clinic, Sex. Trans. Dis. 12:198 (1985).

17. P. A. Granato and M. Roefaro, Comparative evaluation of enzyme immunoassay and culture for the laboratory diagnosis of gonorrhea, Am. J. Clin. Pathol. 83:613 (1985).

18. I. Nachamkin, S. J. Sondheimer, S. Barbargallo and S. Barth, Detection of Neisseria gonorrhoeae in cervical swabs using the Gonozyme enzyme immunoassay, Am. J. Clin. Pathol. 82:461 (1984).

19. J. T. Rudrik, J. M. Waller and E. M. Britt, Efficacy of an enzyme immunoassay with ultracentrifuged first-voided urine for detection of gonorrhea in males, J. Clin. Microbiol. 20:577 (1984).

20. J. Schachter, F. Pang, R. M. Parks, R. F. Smith and A. S. Armstrong, Use of gonozyme on urine sediment for diagnosis of gonorrhea in males, J. Clin. Microbiol. 23:124 (1986).

21. P. A. Totten, K. K. Holmes, H. H. Handsfield, J. S. Knapp, P. L. Perine and S. Falkow, DNA hybridization technique for the detection of Neisseria gonorrhoeae in men with urethritis, J. Inf. Dis. 148:462 (1983).

22. P. L. Perine, P. A. Totten, K. K. Holmes, E. H. Sng, A. V. Ratnam, R.
     Widy-Wersky, H. Nsanze, E. Habte-Gabr and W. G. Westbrook, Evalua-
     tion of a DNA-hybridization method for detection of African and
     Asian strains of Neisseria gonorrhoeae in men with urethritis, J.
     Inf. Dis. 152:59 (1985).
23. A. Janik, E. Juni and G. A. Heyn, Genetic transformation as a tool for
     detection of Neisseria gonorrhoeae, J. Clin. Microbiol. 4:71
     (1976).
24. H. W. Jaffe, S. J. Kraus, T. A. Edwards, S. S. Weinberger and L.
     Zubrzycki, Diagnosis of gonorrhea using a genetic transformation
     test on mailed clinical specimens, J. Inf. Dis. 146:275 (1982).
25. W. L. Whittington, M. Miller, J. Lewis, J. Parker, J. Biddle and L.
     Zubrzycki, Diagnosis of gonococcal infection: A genetic transfor-
     mation test performed on specimens mailed to a clinical laboratory,
     Sex. Trans. Dis., submitted (1987).
26. Centers for Disease Control, "Antibiotic-Resistant Strains of Neisseria
     gonorrhoeae," MMWR (suppl. 5S), vol. 36 (1987).

NUCLEIC ACID HYBRIDIZATION AS A DIAGNOSTIC TOOL

FOR THE DETECTION OF HUMAN PAPILLOMAVIRUSES

Robert F. Rando

Departments of Pathology and OB/GYN
Pennsylvania Hospital
Philadelphia, PA

INTRODUCTION

Human papillomavirus (HPV) infections of the anogenital tract present several difficult questions pertaining to diagnosis. The most obvious question is whether a test for papillomavirus in anogenital lesions is necessary, and if so, what information is desired from such a test. If a diagnostic test is sought, then the next problem concerns the multitude of different papillomavirus types found in anogenital tract lesions, the need, if any, to differentiate these multiple types, and the availability of diagnostic probes to do so. The following discussion will elucidate the arguments supporting the need to test for HPV, by briefly describing a number of epidemiologic studies which associate HPV DNA with premalignant and malignant anogenital tract lesions. Molecular biological data which demonstrates the ability of certain papillomaviruses to transform cells in culture will also be reviewed. In addition, the ways in which nucleic acid hybridization techniques apply to the detection of HPV infections will be described, as well as advantages and disadvantages of several different hybridization techniques which can be used to HPV detection in anogenital lesions.

EPIDEMIOLOGY

Human papillomaviruses (HPVs) are the etiologic agent involved in the induction of condylomas. The presence of HPVs can be determined by analysis of viral nucleic acids (DNA or RNA) and proteins. Using an electron microscope, viral particles (virions) can be regularly demonstrated in the nuclei of condyloma cells. In the 1900's, condylomas were experimentally transmitted from person to person by vaccination with cell-free filtrates[59]. Histologically, similar proliferations have been induced in human foreskin and cervical cells infected with HPV-11 and grafted beneath the renal capsules of athymic mice. Both HPV DNA and group-specific papillomavirus antigens could be demonstrated in the infected grafts[30,31].

HPV infections of the genital tract are one of the most common sexually transmitted viral infections in the United States. Reports from various agencies in the United States and worldwide have shown a dramatic increase in the number of reported cases of anogenital warts. The National Disease and Therapeutic Index conducts an on-going survey for sexually transmitted diseases. This survey is a continuing compilation of statistical information from private practitioners' office-based visits. One such compilation monitored both office and hospital visits for genital warts between 1966 and 1984. The results show a 6.7-fold increase in total visits (169,000 to 1,150,000) and a 4.5-fold increase in first reported cases (53,560 to 224,900). The distribution, by age, for women shows a peak incidence between ages 20 to 24, and for men the peak incidence occurs between 25 to 29. This type of survey must be interpreted with caution

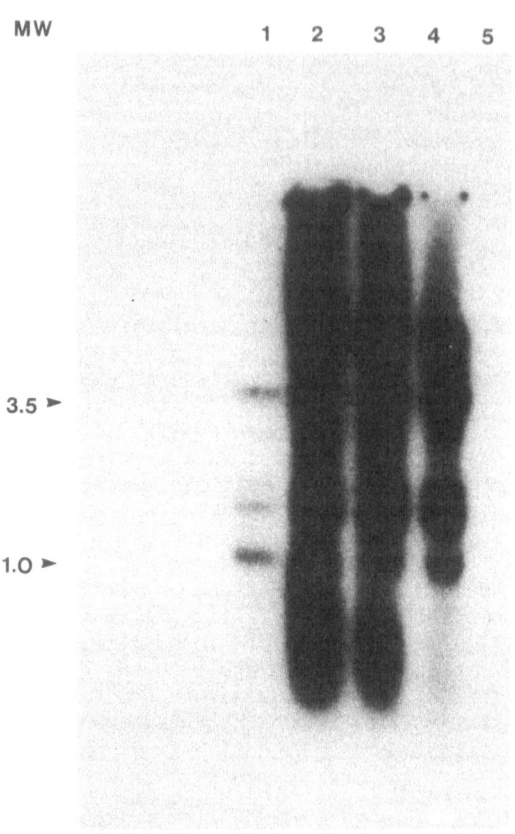

Figure 1. Southern blot analysis of four different samples obtained from the same patient. DNA (10 ug) extracted from a biopsy of the left vagina (lane 1), a biopsy of the cervix at position 2 o'clock (lane 2), 8 o'clock cervical biopsy (lane 3), and exfoliating cervical cells obtained by washing the cervix with sterile saline (lane 4), was digested with the restriction endonulease Pst I, electrophoresed in a 0.8% agarose gel, transferred to a filter membrane and hybridized to ^{32}P-labelled HPV-6 DNA. Lane 5 contained 10 ug of a control DNA. The molecular weight markers (MW) are labelled according to approximate length of nonradioactive DNA marker fragments (in kilobase pairs) which were electrophoresed in lanes to the left of lane 1. This autoratiograph was purposely overexposed to help visualize the faint HPV-6 positive DNA fragments in lane 1.

because the recent media attention on sexually transmitted diseases may increase physician and patient awareness, thus inflating the number of people consulting physicians with this concern. On the other hand, many people with genital warts have been treated in public health clinics and are therefore not mentioned in this survey. Of greater significance is the large number of subclinical HPV infections associated with anogenital lesions[5] which until recently were not recognized as such, and may in fact constitute a majority of HPV infections. In addition, HPV DNA has been detected in 10-20% of biopsies from clinically and histologically normal cervices[40,69].

The standard criteria for the identification of HPV infection has been the clinical observation of exophitic condyloma, or the observation of koilocytotic atypia, akantosis, or dyskeratosis under cytologic or histologic examinations[28,47]. Using the first criterion, cervical condylomata acuminata is detected infrequently. However, the recognition by Meisels that the cellular changes in cervical dysplasia were similar to those seen in condylomas[48,49], was instrumental in reclassification of 90% of mild dysplasia as condyloma planum. This observation, along with the detection of virus particles in crystalline array within the nucleus of a selected koilocytotic cell, in a routine cervical smear, stimulated considerable interest in the association of HPVs with premalignant and malignant lesions of the cervix. The observation that structural antigens of HPV were present in cervical dysplasias[34] and the detection of HPV DNA sequences in these lesions[37], have helped establish the relationship of HPV to cervical neoplasia. Several studies have indicated that 10-20% of women with normal PAP smears and normal colposcopic examinations have HPV DNA in their cervical epithelium[40,74]. It is now generally accepted that most cervical HPV infections are either latent or subclinical, becoming visible only after the application of acetic acid[58]. Subclinical infections of the remainder of the anogenital tract are also apparent using immunocytochemical and DNA hybridization techniques[11]. There is a growing concern over the extent of adjacent tissue infected with HPV when a focus of proliferating cells containing HPV DNA is observed during examination[73]. The extent of viral infections of the anogenital tract is demonstrated in figure 1. Biopsy samples were obtained from different anatomic sites under colposcopic direction after the application of a mild acetic acid solution. There was no exophitic condylomatous growth observed at any sites biopsied, however, the hybridization results demonstrated HPV DNA and the histologic diagnosis of condylomatous atypia, with or without dysplasia, at all of the biopsied sites.

Epidemiologic studies have shown that invasive cancer of the uterine cervix and its precancerous lesions, cervical dysplasia, and carcinoma in situ (cervical intraepithelial neoplasia I to III), are all linked to a sexually transmitted disease. Accumulating data not only associates HPVs with these malignant and premalignant lesions, but also implicates HPVs as the etiologic agent[8,19,20]. Immunocytochemical and molecular deoxyribonucleic acid (DNA) hybridization[37] or ribonucleic acid(RNA) hybridization studies[38,64] have demonstrated the presence of human papillomavirus structural antigens, human papillomavirus nucleic acids, or both, in up to 90% or premalignant and malignant lesions. A causal relationship is strongly suggested by molecular cloning of HPV types 16 and 18 from invasive cervical cancers[8,19] and the subsequent demonstration of these and closely related virus DNAs in the majority of grade three cervical intraepithelial neoplasias (CIN), cervical cancer cell lines[76], and lymph node metastases[36].

It is interesting to note that HPV-16 and HPV-31 show closely related DNA sequences. Both are prevalent in biopsies of various grades of CIN collected from American women[39,57], but differ in their association with cervical cancer[57].

STRUCTURAL AND GENETIC ASPECTS OF PAPILLOMAVIRUSES

Structure of PV capsids and nucleic acids

Papillomaviruses are classified as genus Papillomavirus of the family Papovaviridae[45]. The second genus is comprised of polyoma virus and simian virus 40 (SV40). These genera were grouped together on the basis of similar structures of their viral capsids and nucleic acids. They are distinguished by the size of their naked icosahedral capsids (55 nm vs. 40 nm), and by the molecular weight of their circular, double-stranded DNA genomes (5×10^6 vs 3.3×10^6). Papillomaviruses have been associated with benign, proliferative lesions of skin and other epithelia in a variety of animal species, including humans[52]. While the biology of polyoma and SV40 viruses have been investigated in great detail, the lack of a suitable cell culture system for in vitro propagation of papillomaviruses has severely retarded the investigation of these viruses. Despite this lack of basic biological data, there has been a rapid increases in data covering molecular aspects of this virus. Molecular hybridization techniques have helped identify distantly related HPVs, while molecular cloning technology has led to the isolation of over 50 distinct virus types.

Papillomaviruses are also classified by the species they infect and the degree of DNA cross hybridization between viral genomes (genotypes). At this time, structural antigenic determinants (serotypes) are not used in determining classification. A PV is considered a new type if its DNA genome has less than 50% homology with the genomes of the defined HPV types. This homology criterion refers to 50% hybridization homology and not 50% DNA sequence homology. Each HPV type is usually, but not always, associated with a particular set of clinical or pathologic entities. At least 12 different HPV types have been associated with lesions of the genital tract.

Papillomavirus DNA persists in productive keratinocytes as self-replicating extrachromosomal nuclear episomes. In many higher grade lesions, the nomrmal monomeric closed circular molecules convert to multimeric structures, often an indication of difficulties in DNA replication[72]. In many squamous carcinomas of the uterine cervix associated with HPV infection[4,46,70] and in cervical carcinoma, derived cell lines such as HeLa, SiHa and CaSki[3,51,64,76] HPV DNA can be found integrated into the host genome at one or more locations, often in tandem arrays. It is not known at this time whether integration preceeds transformation or whether it is a causative event in the cellular transformation process. It is noteworthy that integration has occured near cellular oncogenes in several cases studied[17], and there have also been reports of amplified oncogene DNA or mRNA in a significant percentage of cervical cancer[17]. There have also been reports that in certain anogenital tumors associated with HPV types, the viral genomes are not integrated, but instead seem to have developed duplications in the viral replication and transcriptional control regions[9,10,13,53]. In addition, naturally occuring deletion mutants have been observed in several HPV types which can also be found integrated[50,56].

Genomic Organization and Function

The PV genomes are closed-circular, double-stranded DNA molecules consisting of approximately 7500 - 8000 base pairs. The complete DNA sequences of bovine papillomavirus type one (BPV-1), cotton tail rabbit PV type one (CRPV-1), deer fibromavirus type one (DFPV-1), and HPV-1a, 5, 6, 8, 11, 16, 18, and 33 (as well as sequence variation in HPV-6[55]) have been described. Computer aided alignment of these DNA sequences reveal that they all possess a similar genetic organization (Fig. 2). All of the potential gene products encoded in open reading frames (ORFs) are transcribed from the same strand of DNA. The ORFs are arranged in order of size and approxi-

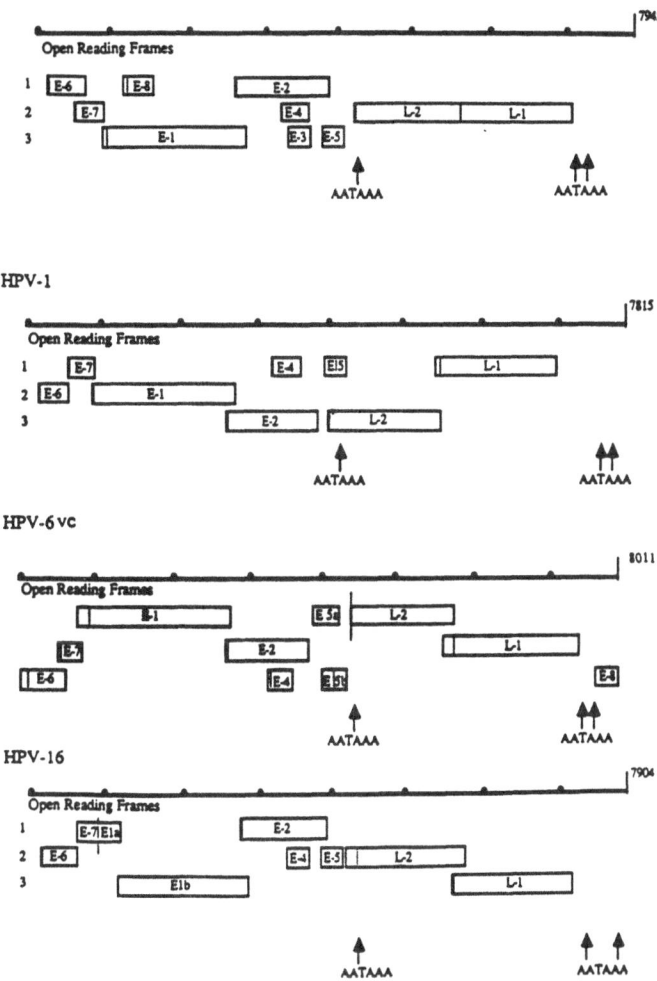

Figure 2. Linear schematic representation of circular papillomavirus genomes. This diagram demostrates the inter- and intra-species conservation
of genetic organization. The open reading frames (ORFs) are organized
according to early "E" or late "L" gene expression and the size of the non-
interrupted (known and putative) protein coding regions. The DNA sequences
between the end of the L1 and beginning of the E6 ORFs contain information
for the control of virus transcription and the origin of viral replication.
The potential mRNA polyadenylation signals AATAAA (and variation of this
motif) are illustrated. The E8 ORF in HPV-6vc is the only example of a
putative protein coding region in this section of the viral genome.

mately in order of temporal expression (early "E" or late "L"). The viral
transcripts utilize more than one promotor[2,14] and undergo extensive post-
transcriptional modification (splicing) of the nascent mRNA chains.

The biological life cycle for a benign papillomavirus infection is
complex and not well understood. It is generally believed that viral infection begins in the basal layers of the epithelium—either cutaneous,
mucosal, or metaplastic. In a natural infection of epithelium, it is likely
that wounding is essential to allow for direct physical contact between
viral particle and the basal epithelium. It seems that either the virus

is latent in these cells or the viral genome is only present in a low copy number, since it is difficult to detect the viral DNA in these cells[52]. The earliest detection of viral DNA or mRNA is in the first or second suprabasal cell layer[22,68]. Viral capsid antigen, which is encoded for by the late ORFs, and mature viral particles are detected as the epithelial layers progress toward terminal differentiation[52]. Evidence is accumulating which describes the role of cell type[55] and cellular differentiation on various aspects of papillomavirus replication and transcription[29,32]. The data describing these events in polyomavirus will help in understanding the same events in papillomaviruses.

Most of the studies investigating particular PV genes has centered around the BPV-1 animal model system. BPV-1 has served as a model for studying PV genetics because it efficiently induces morphological transformation of certain mouse cell lines. In these cells, as in naturally occurring papillomavirus infections, the BPV genome replicates autonomously in the nucleus as a multicopy plasmid. However, as has been found with other PVs, no viral particles are produced in cultured cells, apparently because specific cellular signals required for induction of the vegetative cycle are absent in cultured keratinocytes and fibroblasts. Because of this gap in biological information available through tissue culture, the details of PV vegetative growth are still poorly understood. The genetic analysis of PVs has focused on the aspects of PV biology that can be studied in the transformed mouse fibroblast system, such as the induction of morphologic transformation, autonomous replication of the genome and control of viral early gene expression. The following is a synopsis of what is known about the different regions of the PV genome.

1. <u>Regulatory region</u>. The region of the viral genome which contains the viral replication and trascriptional control elements has been referred to as the upstream regulatory region (URR), the noncoding region (NCR), or long control region (LCR). This area of the biral genome is located between the end of the L1 and the beginning of the E6 ORFs. In addition to the origin of DNA replication[43], this region contains several promotors[3,14,67] and enhancers[55,67] of RNA synthesis. The upstream regulatory region also contains the most DNA sequence divergence differences among the PVs which have been sequenced to date, and some of the differences have been correlated with changes in virulence and oncogenic potential[9,13,53,54,55].

2. <u>Early gene products</u>. The E6 ORF (the sixth largest protein coding region which is transcribed early in an infection. Fig. 2 is clearly involved in cellular transformation and alteration of growth properties of cells in culture. Mutations within the BPV-1 E6 ORF yield only partial transformation of mouse C127 cells[61,62], while the production of E6 protein using a heterologous eukaryotic expression vector in the same mouse cells results in transformation of these cell cultures[63,75].

In the BPV-1-transformed mouse cell system, an E6/E7 fusion protein plays a role in regulating copy number of episomal DNA molecules in each infected cell[7,42]. When this fusion domain is mutated, the average copy number of episomal DNA molecules drops to two or three per cell. In the human cervical carcinomas and derviative cell lines, the E7 region is maintained and expressed as a 20,000 dalton protein which can be phosphorelated[65].

The E1 ORF is directly involved with extra-chromosomal DNA replication[41]. Mutations in the carboxy-terminal two-thirds of the ORF block the ability of the BPV-1 genome to replicate as an episome. After transfection of cell cultures with BPV-1, there is a high rate of viral replication. Eventually the copy number of plasmid DNA drops to the 20-50 per cell characteristic of the steady-state transformed cell culture. BPV-1 recombinant molecules constructed with mutations in the amino-terminal end of the E1 ORF result in loss of such down-regulation[7]. It appears that the E7 or E6/E7 fusion products and the E1 ORF amino-terminal protein product have opposite effects on the maintenance of extrachromosomal copy number of BPV-1 molecules. It is possible that the amino- and carboxy-terminal peptides of the E1 ORF are two distinct proteins derived from the same E1 ORF by the complex mRNA post-transcriptional modifications used by the PVs.

The E2 ORF encodes a <u>trans</u>-acting transcriptional enhancer and a <u>trans</u>-acting transcriptional repressor molecule. The transcriptional enhancer has been described in the BPV-1 and various HPV systems[1,24,67] while the transcriptional repressor has been demonstrated in the BPV-1 system[35]. The two E2 gene products recognize the same DNA sequence ($ACCN_6GGT$), This recognition sequence is conserved not only among different strains of PV, but between species as well. Evidence is accumulating which shows that the PV transcriptional control is also regulated by specific interactions between cellular factors and DNA sequences other than the E2 responsive elements. Rando et al.[55] showed that portions of the HPV-6vc regulatory region, which did not contain the E2 binding motif ($ACCN_6GGT$), could enhance transcription in certain cell lines. Guis and Laimonins[23] have shown an enhancer element in HPV-18 which may be activated by the viral E6 gene product.

The E4 ORF has not yet been described functionally in the BPV-1 transformed cell system or for HPVs even though investigators have described an abundance of HPV-1a mRNAs which map to this region of the viral genome[14] and the E4 protein product of HPV-1a constitutes about 30% of the protein mass of plantar warts[16]. The E4 gene product would offer a tempting target for the production of an immunocytochemical detection system, however this region of the genome is usually deleted when the virus integrates into the host genome in human cancers.

The BPV-1 E5 ORF protein product is part of the viral encoded transforming activity of mouse fibroblast cells in culture[15,21,63]. BPV-1 viral genomes with mutations in the E5 gene are also limited in their ability to replicate autonomously[21]. A synthetic BPV-1 E5 protein has been constructed, using a prokaryotic expression vector, in <u>E. coli</u> and antisera have been raised to this recombinant protein. The antisera were used to localize the naturally occurring BPV-1 protein to the cytoplasmic membrane. The E5 protein is translated from a long mRNA with initiation of protein synthesis starting at an internal initiation codon; this is not typical in eukaryotic cells. With a molecular weight of 7000 daltons, E5 is the smallest viral transforming protein known. The E5 region of the different PV genomes is extremely variable, and is usually deleted in integrated forms of the virus. Consequently, the transformation capabilities of the E5 protein, in both benign and malignant lesions associated with PVs, remains unclear.

There is no information available on the putative E8 ORF found on the BPV-1 and in the viral regulatory region of HPV-6vc (Fig. 2).

3. <u>The Late ORFs</u>. The L1 gene product is the major capsid protein and has a molecular weight of 54,000 daltons. This protein is highly conserved between PVs at both the nucleic acid and protein levels. A polyclonal antibody, developed against disrupted BPV-1 capsids, cross-reacts with most PV L1 proteins. The antibody recognizes a highly conserved regions which is normally not exposed in intact virions, however only about 50% of genital condylomas identified by cytologic features and nucleic acid hybridization analysis produce detectable L1 protein. This percentage decreases as the degree of neoplasia increases[33].

The L2 ORF encodes a minor capsid protein of about 76,000 daltons. This ORF is quite variable among HPVs, and contains a variety of type-specific epitopes which appear to be shielded in intact capsids[27]. The L2 protein is larger than the predicted size based on the translation of the L2 ORF. This difference may be due to post-transcriptional or post-translational modifications[16].

NUCLEIC ACID HYBRIDIZATION TECHNIQUES FOR THE DETECTION OF PAPILLOMAVIRUSES

Principles of hybridization

Deoxyribonucleic acid (DNA) or ribonucleic acid (RNA) molecules can be found as either single-stranded or double-stranded conformations. Intracellular DNA is found as two chains of polynucleotides which exist in a double helical structure. The main chain of each strand consists of deoxyribose residues joined by 3' to 5' phosphodiester covalent bonds. The two chains are held together in part by hydrogen bonding between the base pairs: guanine bases pairing with cytosine bases and thymine bases pairing with adenine bases. In RNA, the thymine bases are replaced with uracil bases which can pair with adenine in either RNA or DNA. The two chains or DNA are coiled to help permit proper base pairing and are anti-parallel to each other. The DNA chains are usually referred to by the direction of the chain, either 5' or 3' end, reading schematically from left to right. A common practice for abbreviating the DNA when describing small segments is to allow an A (adenine), C (guanine), C (cytosine), or T (thymine) signify the entire nucleotide triphosphate with that particular base attached, as shown in the following example.

```
5' --GGATCCGGACATTTAACATCATATTAGCGCGT-- 3'
3' --CCTAGGCCTGTAAATTGTAGTATAATCGCGCA-- 5'
```

In addition to hydrogen bonds, hydrophobic forces between the stacked purine and pyrimidine nuclei contribute significantly to maintenance of the rigid, double-stranded structure. Thus, reagents such as formamide or urea, which increase the solubility of aromatic groups in the surrounding aqueous medium, also tend to destablilize DNA.

Double-stranded DNA can be totally denatured to yield single-stranded complementary molecules. ("Denatured", "dissociated", or "melted" are terms used to describe the separation of the two complementary DNA (or a complementary DNA:RNA) strands.) Denaturation is facilitated by increasing temperature, increasing pH, or by decreasing concentrations of cations. If the double-stranded molecules are already destabilized by the addition of a chaotropic salt or organic solvent, then the ability of temperature, pH, or cation concentration to completely denature the double-stranded molecules is enhanced.

It is remarkable that separated complementary strands of purified DNA recognize each other. Under appropriate conditions they specifically reassociate[44]. The term "hybridization" refers to the formation of stable duplex molecules between two single stranded nucleic acid molecules, whether or not the nucleic acids originated from the same source. However, the term "stable duplex" is relative since both complementary and near complementary single stranded DNA or RNA molecules can form duplexes. Single-stranded molecules which are 100% complementary in their base pair matching will form the most stable duplexes, and annealing or reassociation between single stranded molecules with fewer complementary base pairs form duplexes which are less stable. Advances in the methodology used to form duplexes between related by not 100% complementary single-stranded molecules were instrumental in the detection, isolation, and characterization of over 50 different HPV types.

Hybridization conditions

Most of the data concerning PV detection, isolation, and characterization is a result of hybridization analysis using radiolabeled viral DNA in solution to probe unlabelled DNA immobilized on nitrocellulose filters. The duplex formation in this case is a function of two competing reactions: the reassociation of the radiolabelled DNA $[DNA_s]$, and its hybridization to the unlabelled DNA immobilized on the filter $[DNA_f]$. The initial rate of filter hybridization will be proportional to both $[DNA_s]$ and $[DNA_f]$. When $[DNA_f]$ is much greater than $[DNA_s]$, the contribution of reassociation of $[DNA_s]$ becomes minimal and the rate of hybridization assumes a pseudo-first order characteristic:

$$d[DNA_s]/d(t) = k_1[DNA_f][DNA_s]$$

where $[DNA_f]$ is held constant and the rate constant $[k_1]$ can be solved by experimentally measuring the percentage of bound radioactive DNA at various time points. Factors which would influence the rate of hybridization (Table 1) include the temperature, the concentrations of the two DNA species, the cation concentration, and, to a lesser extent, the length, in base pairs of the DNA fragments involved[26,44]. After the hybridization reactions are finished, the background, or non-specific reactions, are removed by washing the filters with a buffer which is similar in the combined conditions used for hybridization without the radiolabelled probe.

Another parameter which affects the outcome of hybridization reactions is the stability of the hybrids formed. This is particularly important

Table 1. Factors Affecting Hybrid Formation and Stability

Temperature	The optimal rate for duplex formation is T_m $-25°C$
Stringency	The closer the hybridization and wash temperatures are to the T_m, the higher the stringency of the reactions and the % mismatched base pairs formed decreases
[Cation]	Duplex stability and the rate of formation increase with increasing $[Na^+]$ up to approximately 1 molar
[DNA]	Rate of duplex formation increases with increasing [DNA]
pH	Rate of duplex formation decreases as the pH moves farther from neutral. Duplexes denature in strong alkali solutions.
DNA length	Short DNA fragments (200 to 500 base pairs) are optimal for duplex formation.
DNA complexity	Multiple copies of the same DNA sequence will increase the rate of duplex formation while copies of a particular sequence increase the complexity of the DNA mixture and cause a decrease in the rate of duplex formation.
DNA composition	Increasing the percentage of G and C base pairs in a DNA sequence increases the T_m of that sequence.
Base pair mismatch	Duplex stability is reduced by increasing the percentage of mismatched base pairs.

when one wishes to detect distantly related DNA sequences in which the hybrids are not 100% complementary. An example of this is the use of HPV-6 DNA as a molecular probe to detect HPV types 11, 16, 18, etc. The stability of hybrid formation depends on may of the same characteristics as the rate of hybrid formation, with more emphasis placed on the percentage of guanine and cytosine bases in the DNA molecules, and the percentage of mismatched bases which would occur in the annealing of dissimilar DNA sequences. "Stringency" is a term used to describe the degree of discrimination between imperfectly matched and highly matched hybrids through the careful manipulation of DNA concentration, temperature, cation concentration and the concentration of destabilizing factors in the hybridization and washing buffers (Table 1).

The temperature required to completely denature double stranded DNA is called the melting temperature (T_m). The temperature at which most reannealing reactions are conducted, usually in the presence of a standard concentration of sodium ion (1 Molar), is 25°C below T_m (T_m-25°C). This reaction condition is considered moderately stringent. Organic solvents such as formamide have the ability to lower the temperature needed to totally melt double stranded DNA. By taking advantage of this property, one can change the stringency of hybridization by changing the concentration of formamide in the reaction buffers. One can predict the strigency of hybridization using the following reaction: $T_m = 81.5 + 16.6$ (log M) + 0.41 (%G + C) - 0.72 [formamide][26] where M is the monovalent cation concentration (usually 1 molar sodium ion) and %G + C is the percentage of guanine to cytosine residues in the DNA molecules of interest. The farther below the T_m that the hybridizations are conducted, the lower the stringency of the reaction, and the greater the ability of dissimilar DNA moleules to anneal.

Nucleic acid hybridization and HPV diagnosis

The type of hybridization test to be used for the detection of HPV nucleic acids depends on the type of information desired and the availability of the different cloned HPV DNAs associated with the lesions being investigated. For example, HPV-6, 10, 11, 16, 18, 31, 33, 35, and other, as yet uncharacterized virus types, are found associated with the anogenital tract. The following explains several different hybridization techniques and some of the advantages and disadvantages of each type.

1. Southern blot hybridization was first described in 1975[66]. In this procedure, total DNA (cellular and viral) is gently extracted from cells obtained from clinical samples by treatment with a lysis buffer (which may or may not contain a protease), using a modification of the Hirt lysis procedure[25,37]. The DNA samples are then cleaved by digestion with one or more restriction endonucleases and electrophoresed in agarose gels to fractionate the resulting DNA fragments according to size. The resulting pattern of DNA molecules is then superimposed on a nitrocellulose (or more recently, nylon) filter by osmosis as described by Southern[66]. More recently, apparatuses which allow for the electrophoretic transfer from agarose to a solid support have been used.

One of the advantages of Southern blot hybridization can be seen in Figure 3 (lanes b and c), in which two different HPV-6 subtypes were detected in exfoliating cervical cells obtained from two different patients. In this case, the two DNA samples have been cleaved with the restriction endonuclease Pst I, and the resulting banding pattern, after fractionation in 0.8% agarose, shows the loss of a single Pst I site in the HPV-6 viral DNA detected in lane "c" (note arrows 1, 2, and 3 in Fig. 3). The DNA band in lane "c" (arrow 3) corresponds to a summation of the two bands (arrows 1 and 2) in lane "b". This information may be of importance, as

Figure 3. DNA (15 ug) extracts from 10 different samples of exfoliating cervical cells were hybridized to ^{32}P-labelled HPV-6 DNA (panel A) and the identical extracts were hybridized to ^{32}P-labelled HPV-16 DNA (panel B). DNA extracts from samples a, b, c, and d were digested with Pst I and DNA extracts from the remaining samples were digested with BamH I. The DNA of sample "d" was negative for these two papillomaviruses. The molecular weight (MW) determinations were made as described in Figure 1.

several investigators have desribed particular HPV-6 subtypes (identified by their DNA banding pattern after cleavage with particular restriction endonuclease) to be associated with malignancy[9,10,53,54].

In order to accurately determine the type of HPV involved with a particular lesion, the proper HPV DNA clones (which are used as molecular probes) need to be available. In Figure 3, the DNA extracts in panel A were hybridized to ^{32}P-labelled HPV-6 DNA, and the identical extracts, in panel B, were hybridized to ^{32}P-labelled HPV-16 DNA. The viral DNA in

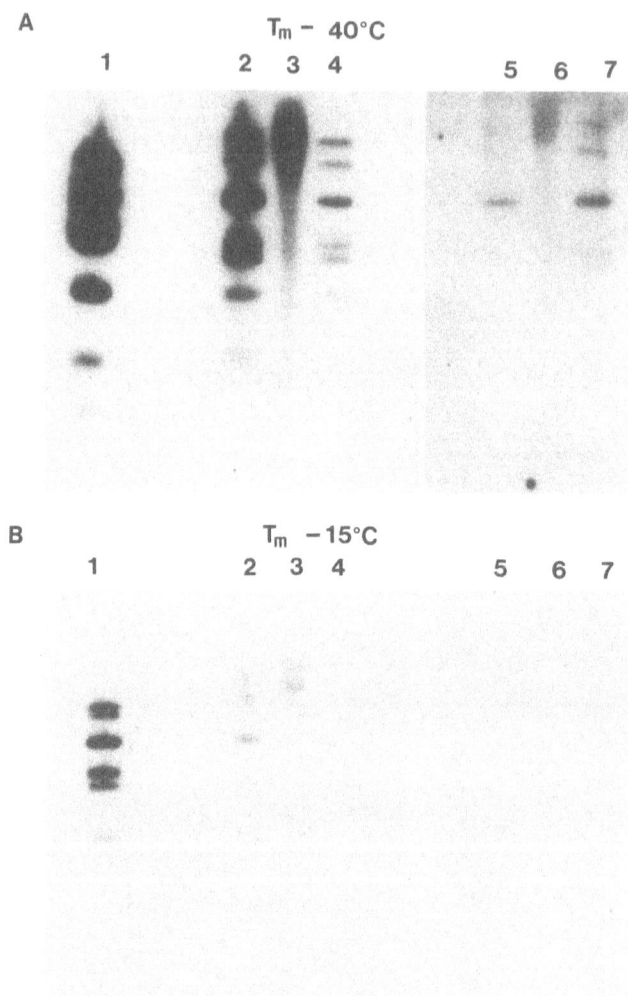

Figure 4. DNA extracts (10ug) from biopsy specimens were hybridized to ^{32}P-labelled HPV-16 DNA (lanes 2, 3, 4) or the identical extracts were hybridized to ^{32}P-labelled HPV-6 DNA (lanes 5, 6, 7). All of the samples (lanes 2-7) were digested with Pst I before electrophoreses and transfer to filter membranes. The filter membranes were hybridized and washed at T_m -40°C (panel A) and then rewashed at T_m -15°C (panel B). Lane 1 contains 20 ng of HPV-16 which has been inserted into the bacterial plasmid vector pBR322 at the unique BamH I site. This control DNA was digested with Pst I before electrophoresis. The autoradiographic exposures for both wash conditions were for 2 days at -70°C.

lanes "a" and "e" hybridized much better to the HPV-16 probe (panel B) and
the viral DNA in lanes "b", "c", "f", "h", "i", and "j" formed hybrids
better with the HPV-6 probe (panel A). One can note that the standard
conditions used in these hybridizations (T_m - 25°C) allowed for a slight
degree of cross hybridization with the DNA in all lanes, indicating some
degree of DNA sequence similarity between the probes and the HPV DNA present in all of the samples. The ability to detect HPVs which are more distantly related than HPV-16 and HPV-6 increases as one decreases the stringency of the reaction, as can be seen in Figure 4. In this experiment,
faintly hybridizing band of HPV DNA which were visible when the hybridization
and washing steps were performed at T_m - 40°C were no longer seen when the
filters were rewashed at T_m - 15°C. Hybridization and washing conditions
at low stringency are sometimes hard to interpret due to background hybridization (a non-specific attachment of radiolabelled nucleotides to the
filter membrane or to the cellular DNA). The ability to distinguish this
"background" from specific viral hybridization can be accomplished by using
the Southern blot technique.

Other information which can also be obtained only from Southern blot
analysis is the presence of viral DNA fragments which cannot be explained
by the predicted number and size of DNA fragments produced when episomal
viral DNA is digested with particular restriction endonucleases. Aberrant
DNA bands can be detected when the viral genome integrates into the host
genome (which is observed in the higher grades of dysplasia and in carcinomas associated with HPV-16 and HPV-18[8,18,38]). Another explanation for
aberrant bands in the Southern blot analysis is the presence of naturally
occurring deletion mutants which may be precursors to, or byproducts of,
viral integration. An example of this type of band is shown in Figure 3,
panel B (arrow 4). By isolation and characterization of this DNA fragment,
this band has been confirmed to be a naturally occurring deletion mutation
of HPV-16. The truncated DNA genome has been analysed by restriction endonuclease mapping and DNA sequencing using the dideoxy chain termination
technique[60]. The results of this analysis (Fig. 5) located two distinct
mutations in the HPV-16 genome. One of the mutations resulted in the loss
of most of the L2 ORF, and the second mutation resulted in the loss of
sequences from the 3' end of the viral control region up to the 5' end of
the E1 ORF.

The major drawbacks to the use of Southern blot analysis are the technical complexity of the assay, the amount of time needed for analysis, and
the use of radiolabelled compounds. In addition, it is not possible using
this technique to ascertain the exact anatomic site of infection, whether
the infection is multifocal in the lesion in question, and whether the
tumor is actually involved with the HPV DNA or whether the HPV infection
is in a site adjacent to the tumor.

2.   Dot blot hybridization is similar to Southern blot hybridization,
except that rather than digesting the extracted DNA with restriction endonucleases and fractionating the agarose gels, the extracted DNA is simply
dotted onto nitrocellulose membranes. This process is faster and less
costly than the Southern blot procedure, and, depending on the type of
information one requires it may be more advantageous. One of the drawbacks
to this procedure is the inability to perform non-stringent hybridization,
due to the difficulty in the interpretation of a non-specific hybridization
from weakly hybridizing HPV DNA sequences. These signals can be distinguished using the Southern blot technique (Fig. 5). Therefore, one must
acquire all the HPV DNA probes for which one is interested in screening.
Another drawback to this procedure is the inability to distinguish HPV
subtypes (because no restriction endo nuclease analysis is performed) or
to distinguish HPV DNA which has become integrated from episomal HPV DNA.
One assumption which is made when describing both Southern blotting and

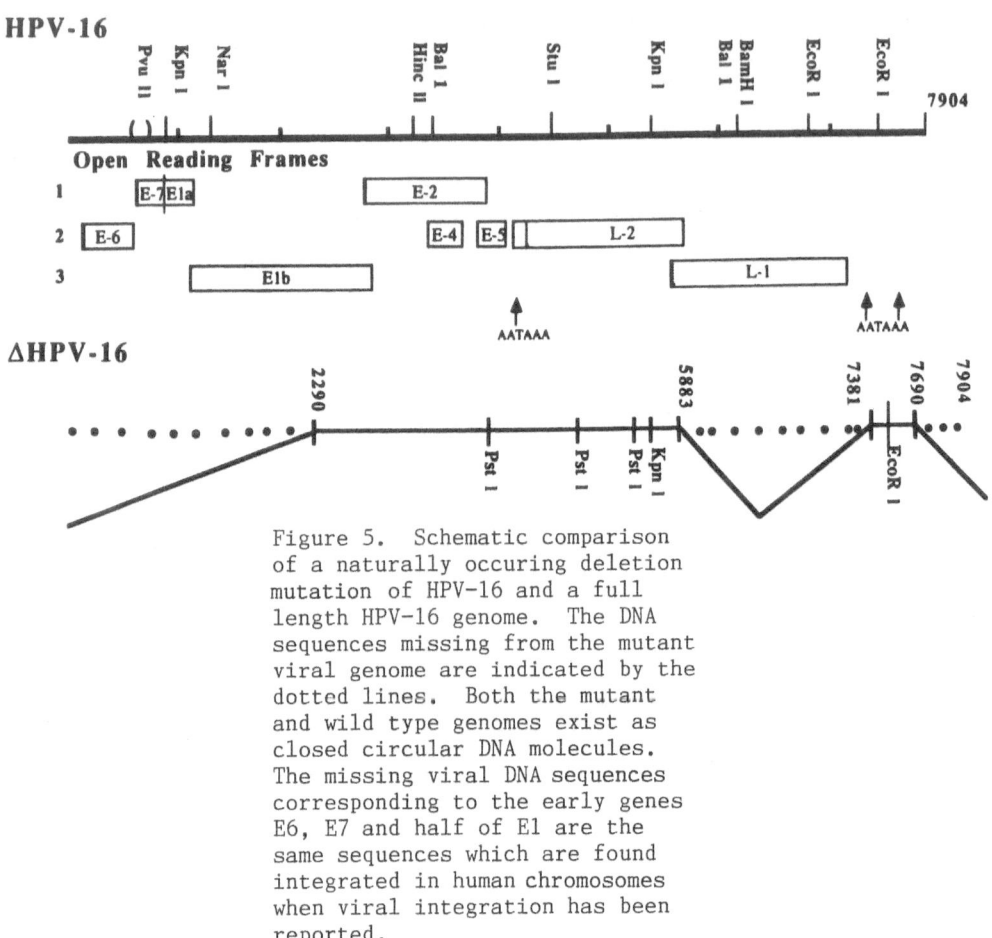

Figure 5. Schematic comparison
of a naturally occuring deletion
mutation of HPV-16 and a full
length HPV-16 genome. The DNA
sequences missing from the mutant
viral genome are indicated by the
dotted lines. Both the mutant
and wild type genomes exist as
closed circular DNA molecules.
The missing viral DNA sequences
corresponding to the early genes
E6, E7 and half of E1 are the
same sequences which are found
integrated in human chromosomes
when viral integration has been
reported.

dot blotting is that at some point in the analysis, the actual quantitation
of DNA applied to the filter membranes has been determined.

3.   Filter In-Situ Hybridization. This technique was developed[71] as a
rapid test for the presence of HPV DNA in exfoliating cervical cells which
are obtained by either physically scraping the cervix or by washing the
exfoliating cells off of the cervix using a sterile solution. It is hard
to modify this test for use with any other tissue collection method such
as biopsy material. The clinical samples are suspended in saline to prevent clumping, filtered onto a nitrocellulose (or nylon) membrane, and
the intracellular DNA is denatured, while stuck on the filter membrane, in
a strong alkali solution. The filters are then neutralized and baked in
order to fix the DNA samples. Hybridizations and washes are performed
essentially the same as with other techniques. The advantages of this
system are the speed and ability to screen a large number of samples. The
disadvantages of this system as compared to the Southern blot procedure
are the same as with dot blotting. Filter in-situ hybridizations may be
even less sensitive than dot blot and more prone to false positives due to
non-specific interactions of the radiolabelled probe with cellular debris.

4.   In-Situ Hybridizations. In-Situ hybridization can be performed on
either cytologic cell preparations such as PAP smears or on routinely processed formalin-fixed parafin-embedded biopsy specimens. The major advant-

ages of using in-situ hybridization (over any of the hybridization techniques mentioned above) is in the type of information which can be obtained from this technique. To date there is not an adequate tissue culture system for the study of PVs; this results in a large gap in the understanding of the biology of these viruses. Only recently has an animal model system been developed[30,31] which can help in answering some questions. In-situ hybridization, the only nucleic acid hybridization technique which does not destroy cellular morphology of the biopsied specimens, is ideally suited to answer some of the questions regarding the basic biology and molecular biology of these viruses. One can obtain information concerning the distribution of viral DNA in infected tissue (disseminated or focal), a quantitative measure of the copy number of viral genomes in different cells, the differentiated state of the epithelial cell infected, and whether the infection is actually involved with malignant tumors or is just adjacent to them. In addition, a rough determination of the the temporal expression of viral mRNA transcrips can be determined[68] in both benign HPV infections and in those infections associated with malignancies. Another advantage of the in-situ method is the ability to perform retrospective studies using archival histology sections.

Details of cellular structure can be maintained only if a low energy emitting isotope such as ^{3}H[68] or an enzymatic conjugated system such as avidin-biotin-peroxidase[6] is used. The problem with using ^{3}H as a radioactive label is that the exposure time needed to detect the signal can be as much as one month[68]. The problem with using an enzyme linked colorometric assay is that the sensitivity is usually decreased by almost one order of magnitude relative to a radioactively labelled probe. The use of higher energy emitting isotopes will give a faster signal, but obscure the cellular morphologic details. In-situ techniques can screen a large number of samples simultaneously; however, it is hard to interpret the results of non-stringent hybridization using this technique. This necessitates the acquisition of a large number of different HPV DNA probes, or the limitation of studies to those lesions which contain HPV DNA for which one has a probe. The need to obtain a large number of HPV molecular probes is necessary if one wishes to use this technique as a screening test for the presence of HPV infections in biopsy material.

SUMMARY

At this time nucleic acid hybridization tests are the most sensitive and reproduceable methods for the detection and differentiation of HPV types in clinical samples. The hybridization method of choice depends on the information desired and the availability of the proper diagnostic nucleic acid probes. Assuming most of the HPV nuleic acid probes become readily available in the near future, then the most sensitive test for screening clinical specimens--albeit the most laborious--will be the Southern blot procedure. As more information covering the involvement of HPV infections with the progression of lesions from benign to malignant is compiled, the need to know the particular subtype or status of HPV integration may become more or less important in the screening of clinical samples. If this information becomes less important, or unnecessary for a simple screening procedure, then dot blot hybridization may prove to be a much easier method for obtaining the information desired. If the sensitivity of in-situ hybridization using non-radioactive probes increases, then this method would become the fastest, easiest, and cleanest technique for the screening of a large number of clinical samples where limited information is desired. The ideal test for the future would be automated. In order to automate a test for HPV infections, the test design must become much simpler, and in order to design a simpler test, more information will be needed concerning the biology of HPV infections, how they cause benign or malignant cellular proliferation, and how the host immune system responds to HPV infections.

REFERENCES

1. Androphy, E.J., D.R. Lowry, and J.T. Schiller, Bovine papillomavirus E2 transactivating gene product binds to specific sites in papillomavirus DNA, Nature. 325:70-73 (1987).

2. Baker, C.C., and P.M. Howley, Differential promotor utilization by the bovine papillomavirus in transformed cells and productively infected wart tissued, EMBO J. 6:1027-1035 (1987).

3. Baker, C.C., W.C. Phelps, V. Lindgren, M.J. Braun, M.A. Gonda, and P.M. Howley, Structural and transcriptional analysis of human papillomavirus type 16 sequences in cervical carcinoma cell lines, J. Virol. 61:962-971 (1987).

4. Beaudenon, S., D. Kremsdorf, O. Croissant, S. Jablonska, S. Wain-Hobson, and G. Orth, A novel type of human papillomavirus associated with genital neoplasia, Nature. 321:246-249 (1986).

5. Becker, T.M., K.M. Stone, and E.R. Alexander, Genital human papillomaviurs Infection: A growing concern, p. 389-396, in: "Obstetrics and Gynecology Clinics of North America," R. Reid, ed., vol. 14:2, W.B. Saunders Co., Philadelphia (1987).

6. Beckmann, A.M., D. Myerson, J.R. Daling, N.B. Kiviat, C.M. Fenoglio, and J.K. McDougall, Detection and localization of human papillomavirus DNA in human genital condylomas by in situ hybridization with biotinylated probes, J. Med. Virol. 16:265-273 (1985).

7. Berg, L.J., K. Singh, and M. Botchan, Complementation of a bovine papillomavirus low-copy-number mutant: Evidence for a temporal requirement of the complementing gene, Mol. Cell. Biol. 6:859-869 (1986).

8. Boshardt, M., L. Gissmann, H. Ikenberg, A. Kleinheinz, W. Schewlen, and H. zur Hausen, A new type of papillomavirus DNA, its presence in genital cancer biopsies and in cell lines derived from cervical cancer, EMBO 3:1151-1157 (1984).

9. Boshart, M., and H. zur Hausen, Human papillomaviruses in Buschke-Lowenstein tumors: Physical state of the DNA and identification of a tandem duplication in the noncoding region of a human papillomavirus 6 subtype, J. Virol. 58:963-966 (1986).

10. Byrne, J.C., M.S. Tsao, R.S. Fraser, and P.M. Howley, Human papillomavirus-11 DNA in a patient with chronic laryngotracheobronchial papillomatosis and metastatic squamous-cell carcinoma of the lung, New England J. of Med. 317:873-878 (1987).

11. Campion, M.J., Clinical manifestations and natural history of genital human papillomavirus infections, p. 363-388, in: "Obstetrics and Gynecology Clinics of North America," R. Reid, ed., vol. 14:2, W.B. Saunders Co., Philadelphia (1987).

12. Cole, S.T., and O. Danos, Nuclotide sequence and comparative analysis of the human papillomavirus type 18 genome: Phylogeny of papillomaviruses and repeated structure of the E6 and E7 gene products, J. Mol. Biol. 193:599-608 (1987).

13. Cole, S.T., and R.E. Streek, Genome organization and nucleotide sequence in human paillomavirus type 33, which is associated with cervical cancer, J. Virol. 58:991-995 (1986).

14. Chow, L.T., M. Nasseri, S.M. Wolinsky, and T.R. Broker, Human papillomavirus types 6 and 11 mRNAs from genital condylomata acuminata, J. Virol. 61:2581-2583 (1987).

15. DiMaio, D., D. Guralski, and J.T. Schiller, Translation of open reading frame E5 of bovine paillomavirus is required for its transforming activity, Proc. Natl. Acad. Sci. USA 83:1797-1801 (1986).

16. Doorbar, J., D. Campbell, R.J. Grand, and P.H. Gallimore, Identification of the human papillomavirus-1a E4 gene products, EMBO J. 5:355-362 (1986).

17. Durst, M., C.M. Croce, L. Gissmann, et al., Papillomavirus sequences integrate near cellular oncogenes in some cervical carcinomas, Proc. Natl. Acad. Sci. USA 84:1070-1074 (1987).

18. Durst, M., A. Kleinheinz, M. Hotz, and L. Gissmann, The physical state of human papillomavirus type 16 DNA in benign and malignant tumors, J. Gen. Virol. 66:1515-1522 (1985).

19. Durst, M., L. Gissmann, H. Ikenberg, and H. zur Hausen, A papillomavirus DNA from a cervical carcinoma and its prevalence in cancer biopsies from different geographic regions, Proc. Natl. Acad. Sci. USA 80:3812-3815 (1983).

20. Gissmann, L., M. Boshart, M. Durst, H. Ikenberg, D. Wagner, and H. zur Hausen, Presence of human papillomavirus (HPV) in genital tumors, J. Invest. Dermatol. 83:265-285 (1984).

21. Groff, D.E., and W.D. Lancaster, Genetic analysis of the 3' early region transformation and replication functions of bovine papillomavirus type 1, Virol . 150:221-230 (1986).

22. Grussendorf, E.I., and H. zur Hausen, Localization of viral DNA replication in sections of human warts by nucleic acid hybridization with complementary RNA of human papillomavirus type 1, Arch. Dermatol. Res. 264:55-63(1979).

23. Guis, D., and L. Laimins, Constitutive and inducible expression of the HPV-18 enhancer, p. 126, in: "Sixth International Papillomavirus Workshop," W.D. Lancaster and A.B. Jenson, eds., Georgetown University (1987).

24. Hirochika, H., T.R. Broker, and L.T. Chow, Enhancers and trans-acting E2 transcriptional factors of papillomaviruses, J. Virol. 61:2599-2606 (1987).

25. Hirt, B., Selective extraction of polyoma DNA from infected mouse cell cultures, J. Mol. Biol. 26:365-369 (1967).

26. Howley, P.M., M.A. Israil, M.F. Law, and M.A. Martin, A rapid method for detecting and mapping homology between heterologous DNAs, J. Biol. Chem. 254:4876-4883 (1979).

27. Komly, C.A., F. Breitburd, O. Croissant, and R.E. Streek, The L2 open reading frame of human papillomavirus type 1a encodes a minor structural protein carrying type specific antigens, J. Virol. 60:813-816 (1986).

28. Koss, L.G., and G.R. Durfee, Unusual patterns of squamous epithelium of the uterine cervix: cytologic and pathologic study of koilocytotic atypia, Ann. New York Acad. Sci. 63:1245 (1956).

29. Kovesdi, I., M. Satake, K. Furukawa, R. Reichel, Y. Ito, and J.R. Nevins, A factor discriminating between the wildtype and a mutant polyomavirus enhancer, Nature 328:87-89 (1987).

30. Kreider, J.W., M.K. Howett, N.L. Lill, G.L. Bartlett, R.J. Zaino, T.V. Sedlacek, and R. Mortel, In vitro transformation of human skin with human papillomavirus type 11 from condylomata acuminata, J. Virol. 59:369-376 (1986).

31. Kreider, J.W., M.K. Howett, S.A. Wolfe, G.L. Bartlett, R.J. Zaino, T.V. Sedlacek, and R. Mortel, Morphological transformation of human uterine cervix with papillomavirus from condylomata acuminata, Nature. 317:639-640 (1985).

32. Kryszke, M.H., P. Jacques, and M. Yaniv, Induction of a factor that binds to the polyomavirus A enhancer on differentiation of embryonal carcinoma cells, Nature. 328:254-256 (1987).

33. Kurman, R.J., A.B. Jenson, and W.D. Lancaster, Papillomavirus infection of the cervix: 2. Relationship to intraepithelial neoplasia based on the presence of specific viral structural proteins, Am. J. Surg. Path. 7:39-52 (1983).

34. Kurman, R.J., K.H. Shah, W.D. Lancaster, and A.B. Jenson, Immunoperoxidase localization of papillomavirus antigens in cervical dysplasia and vulvar condylomas, Am. J. Obstet. Gynecol. 140:931-935 (1981).

35. Lambert, P.F., B.A. Spalholz, and P.M. Howley, A transcriptional repressor encoded by BPV-1 shares a common carboxy-terminal domain with the E2 transactivator, Cell. 50:69-78 (1987).

36. Lancaster, W.D., C. Castellano, C. Santos, G. Delgado, R.J. Kurman, and A.B. Jenson, Human papillomavirus deoxyribonucleic acid in cervical carcinoma from primary and metastatic sites, Am. J. Obstet. Gynecol. 154:115-119 (1986).

37. Lancaster, W.D., R.J. Kurman, L.E. Sang, S. Perry, and A.B. Jenson, Human papillomavirus: detection of viral DNA sequences and evidence for molecular heterogeneity in metaplasias and dysplasias of the uterine cervix, Intervirol. 20:202-212 (1983).

38. Lehn, H., P. Krieg, and G. Sauer, Papillomavirus genomes in human cervical tumors: Analysis of their transcriptional activity, Proc. Natl. Acad. Sci. USA 82:5540-5544 (1985).

39. Lorincz, A.T., W.D. Lancaster, and G.F. Temple, Cloning and characterization of the DNA of a new human papillomavirus from a woman with dysplasia of the uterine cervix, J. Virol. 58:225-229 (1986).

40. Lorincz, A.T., G.F. Temple, J.A. Patterson, A.B. Jenson, and W.D. Lancaster, Correlation of cellular atypia and human papillomavirus deoxyribonucleic acid sequences in exfoliated cells of the uterine cervix, Obstet. Gynecol. 68:508-512 (1986).

41. Lusky, M., and M.R. Botchan, Genetic analysis of bovine papillomavirus type 1 transacting replication factors, J. Virol. 53:955-965 (1985).

42. Lusky, M., and M.R. Botchan, Transient replication of bovine papil-
     lomavirus type 1 plasmids: cis and trans requirements, Proc. Natl.
     Acad. Sci. USA 83:3609-3613 (1986).

43. Lyon, J., L. Rosl, and H. Zentgraf, Origin of replication in episomal
     bovine papillomavirus type 1 DNA isolated from transformed cells,
     EMBO J. 3:2173-2178 (1984).

44. Marmur, J., R. Rownd, and C.L. Schildkraut, Prog. Nucleic Acid Res.
     1:231 (1963).

45. Mathews, R.E., Classification and nomenclature of viruses, Intervirol.
     17:1-199 (1982).

46. Matsukura, T., T. Kanda, A. Furuno, H. Yoshikawa, T. Kawana, and K.
     Yoshike, Cloning of monomeric human papillomavirus type 16 DNA inte-
     greated within cell DNA from a cervical carcinoma, J. Virol. 58:979-
     982 (1986).

47. Meisels, A., and R. Fortin, Condylomatous lesions of the cervix and
     vagina, I. Cytologic patterns, Acta. Cytol. 20:505-509 (1976).

48. Meisels, A., R. Fortin, and M. Roy, Condylomatous lesions of the
     cervis, II. Cytologic, colposcopic and histopathologic study, Acta.
     Cytol. 21:379-390 (1977).

49. Meisels, A., C. Morin, and M. Casas-Cordero, Human papillomavirus
     infection of the uterine cervix, Int. J. Gynecol. Path. 1:75-94 (1982).

50. Ostrow, R.S., K.R. Zachow, and A.J. Faras, Molecular coning and nucleo-
     tide sequence analysis of several naturally occurring HPV-5 deletion
     mutant genomes, Virol. 158:235-238 (1987).

51. Pater, M.M., and A. Pater, Human papillomavirus types 16 and 18
     sequences in carcinoma cell lines, Virol. 145:313-318 (1985).

52. Pfister, H., Biology and biochemistry of papillomaviruses, Rev. Physiol.
     Biochem. Pharmacol. 99:111-181 (1984).

53. Rando, R.F., D.E. Groff, J.G. Chirikjian, and W.D. Lancaster, Isolation
     and characterization of a novel human papillomavirus type 6 DNA from
     an invasive vulvar carcinoma, J. Virol. 57:353-356 (1986).

54. Rando, R.F., T.V. Sedlacek, J. Hunt, A.B. Jenson, R.J. Kurman, and
     W.D. Lancaster, Verrucous carcinoma of the vulva associated with an
     unusual type 6 human papillomavirus, Obstet. and Gynecol. 67:70s-
     75s (1986).

55. Rando, R.F., W.D. Lancaster, P. Han, and C. Lopez, The noncoding
     region of HPV-6vc contains two distinct transcriptional enhancing
     elements, Virol. 155:545-556 (1986).

56. Rando, R.F., Naturally occuring deletion mutants of HPV-16 isolated
     from a precancerous vaginal lesion, Virol. submitted (1987).

57. Reid, R., M. Greenberg, A.B. Jenson, M. Husain, J. Willett, Y. Daoud,
     G. Temple, and A.T. Lorincz, Sexually transmitted paillomaviral
     infections. I. The anatomic distribution and pathologic grad of
     neoplastic lesions associated with different viral types, Am. J.
     Obstet. Gynecol. 156:212-222 (1987).

58. Reid, R., C.R. Laverty, M. Coppleson, W. Isarangkul, and E. Hills, Noncondylomatous cervical wart virus infection, Obstet. Gynecol. 55: 476-483 (1980).

59. Rowson, K.E.K., and B.W.J. Mahy, Humanpapova (wart) virus, Bacteriol. Rev. 31:100 (1967).

60. Sanger, F., and A.R. Coulson, A rapid method for determining sequences in DNA by primed synthesis with DNA polymerase, J. Mol. Biol. 94:441-448 (1975).

61. Sarver, N., M.S. Rabson, Y.C. Yang, J.C. Byrne, and P.M. Howley, Localization and analysis of bovine papillomavirus type 1 transforming functions, J. Virol. 52:377-388 (1984).

62. Schiller, J.T., W.C. Vass, and D.R. Lowry, Identification of a second transforming region in bovine papillomavirus DNA, Proc. Natl. Acad. Sci. USA 81:7880-7884 (1984).

63. Schiller, J.T., W.C. Vass, K.H. Vousdan, and D.R. Lowy, The E5 open reading frame of bovine papillomavirus type 1 encodes a transforming gene, J. Virol. 57:1-6 (1986).

64. Schwarz, E., U.K. Freese, L. Gissmann, W. Mayer, B. Roggenbuck, A. Stremlau, and H. zur Hausen, Structure and transcription of human papillomavirus sequences in cervical carcinoma cells, Nature. 314: 111-114 (1985).

65. Smotkin, D., and F.O. Wettstein, The major human papillomavirus protein in cervical cancers is a cytoplasmic phosphoprotein, J. Virol. 61:1686-1689 (1987).

66. Southern, E.M., Detection of specific sequences among DNA fragments separated by gel electrophoresis, J. Mol. Biol. 98:503-517 (1975).

67. Spalholz, B.A., Y.C. Yang, and P.M. Howley, Transactivation of a bovine papillomavirus transcriptional regulatory element by the E2 gene product, Cell. 42:183-191 (1985).

68. Stoler, M.H., and T.R. Broker, In situ hybridization detection of human papillomavirus DNA and messenger RNA in genital condylomas and a cervical carcinoma, Hum. Path. 17:1250-1258 (1986).

69. Toon, P.G., J.R. Arrand, L.P. Wilson, and D.S. Sharp, Human papilloma-virus infection of the uterine cervix of women without cytological signs of neoplasia, Br. Med. J. 293:1261-1264 (1986).

70. Tsunokawa, Y., N. Takebe, S. Nozawa, T. Kasamatsu, L. Gissman, H. zur Hausen, M. Terada, and T. Sugimura, Presence of human papilloma-virus type-16 and type-18 DNA sequences and their expression in cervical cancers and cell lines from Japanese patients, Int. J. Cancer. 37:499-503 (1986).

71. Wagner, D., H. Ikenberg, N. Boehm, and L. Gissmann, Identification of human papillomavirus in cervical swabs by deoxyribonucleic acid in situ hybridization, Obstet. Gynecol. 64:767-772 (1984).

72. Wettstein, F.O., and J.G. Stevens, Variable-sized free episomes of shope papillomavirus DNA are present in all non-virus producing neo-plasms and integrated episomes are detected in some, Proc. Natl. Acad. Sci. USA 79:790-794 (1982).

73. Wickenden, C., A.D. Malcolm, A. Steele, and D.V. Coleman, Screening for wart virus infection in normal and abnormal cervices by DNA hybridization of cervical scrapes, Lancet. 1:65-67 (1985).

74. Winkler, B., V. Capa, W. Reumann, A. Ma, R. La Porta, S. Reilly, P.M. Green, R.M. Richart, and C.P. Crum, Human papillomavirus infection of the esophagus. A clinicopathologic study with demonstration of papillomaviurs antigen by the immunoperoxidase technique, Cancer. 55:149-155 (1985).

75. Yang, Y.C., B.A. Spalholz, M.S. Rabson, and P.M. Howley, Dissociation of transforming and trans-activation functions for bovine papilloma-virus type 1, Nature. 318:575-577 (1985).

76. Yee, C., I. Krishnan-Hewlett, C.C. Baker, R. Schlegel, and P.M. Howley, Presence and expression of HPV sequences in human cervical carcinoma cell lines, Am. J. Path. 119:361-366 (1985).

THE RAPID DETECTION OF CLINICALLY SIGNIFICANT BACTERIA

Richard C. Tilton

Professor of Laboratory Medicine
Department of Laboratory Medicine
University of Connecticut School of Medicine
Farmington, CT 06032

Much has changed in the clinical laboratory. Until 1985, there was attention paid to hospital census and the role of inpatient service in the provision of health care. Naturally, the clinical laboratory contributed much, usually as a "cash cow." Since 1985, there has been a growing dependence on ambulatory services. Not only have hospital outpatient clinics grown but there has been an expansion of day treatment and day surgery. Because of the fiscal necessity for shortened inpatient stays, a substantial part of the health care dollar has been taken up by extended care facilities. The clinical laboratory, then, is competing for a smaller share of the available resources.

Contemporary laboratory economists state that since 1985 laboratories have become fiscal millstones because they are "revenue negative." To compound matters, the tests that were usually thought of as money makers such as the CBC, urinalysis, urine cultures, and throat cultures, are moving inexorably into the physician office setting.

For the microbiologist, then, the choice of methods for detection of bacteria becomes a critical one which must be based on both clinical effectiveness and cost effectiveness.

The issue of rapidity in microbiology testing has been dealt with in many symposia and can best be summarized by indicating that while there may be little objective evidence for the advantages of rapid testing, it is implicit that if a function can be done faster with similar cost and accuracy, then such a procedure is desirable. The issue of rapid testing versus traditional overnight cultural microbiology will not be discussed in this essay. Rather, the equally important question of "why determine etiology at all" must be raised. Ellner (1987) cites 3 reasons to determine etiology. The first is that etiology of a bacterial isolate may be the first clue to underlying disease. Ellner uses the example of the relationship between Streptococcus bovis bacteremia and carcinoma of the colon. He also states that a number of organisms may be involved in the same syndrome. For example, primary atypical pneumonia can be caused not only by Mycoplasma but by Chlamydia and a variety of treatable viruses. Lastly, the source of infection may be suggested by the nature of the isolate; that is, environmental, gastrointestinal or respiratory, nosocomial, food borne, etc.

Rapid Methods in Clinical Microbiology
Edited by B. Kleger *et al.*
Plenum Press, New York

A wide variety of techniques exist for the detection of bacteria. These include light and fluorescent microscopy, immunologic, and molecular methods. The Gram stain continues to be a most effective rapid and accurate method for diagnosis of infection. For example, the specificity of the Gram stain for Haemophilus influenzae and Neisseria meningitidis in cerebrospinal fluid is greater than 98%. Doern (1985) indicates that the Gram stain may be the only clue available for the diagnosis of Branhamella or Neisseria pneumonia. Gonorrhea diagnosis by the Gram stain in symptomatic males continues to be > 95% sensitive.

It is not the intention of the author to describe in detail the various immunologic and molecular methods that are available for the detection of bacteria. Rather, a number of critical statements can be made using these innovative methods of diagnosis as examples. The statements include a) the clinical relevance of antigen detection tests, b) the statistical relevance of antigen detection tests, and c) modification of traditional diagnostic and therapeutic modalities by the availability of new tests for detection of bacteria.

Clinical relevance of antigen detection tests

For the past decade, the detection of bacterial antigens in cerebrospinal fluid (CSF) has been widely acclaimed as useful. Tilton et al (1984) and others have compared sensitivity and specificity of a wide variety of latex agglutination and coagglutination tests for the rapid detection of H. influenzae, N. meningitidis, Streptococcus pneumoniae and Group B streptococcus antigens directly in CSF. Until recently, it has been assumed that these tests were useful in managing a patient with meningitis. However, some investigators have cast doubt on the clinical relevancy of such tests. Gerber (1985) indicated that the examination of CSF in a child suspected of meningitis must answer two questions. 1) Does the patient have bacterial meningitis?, and 2) What is the etiology? The answer to the question "Does the patient have bacterial meningitis?" is aided by a number of laboratory tests including the differential cell count in the spinal fluid, the ratio of CSF glucose to protein and the Gram stain. Gerber (1985) further indicated that 95% of CSF specimens will have normal values. Approximately 5%, then, may benefit from antigen detection. He also addressed the issue that the detection of antigen in the spinal fluid does not provide a susceptibility report. The susceptibility of Haemophilus influenzae to ampicillin especially is not predictable. Empiric therapy has to be instituted in any event. Antigen detection is no more sensitive than the Gram stain. There is a growing unwillingness to depend on antigen detection tests unless it can be shown that antigen detection is both 100% sensitive and 100% specific. While not appreciated by non-microbiologists, the use of the Gram stain in untrained hands also lacks specificity. The conclusion, then, suggests that antigen detection tests for the diagnosis of meningitis may be clinically useful but their utility is limited to the confirmation of the Gram stain, use in antibiotic treated patients or to distinguish between bacterial and viral infection when the usual diagnostic tests on cerebrospinal fluid are negative and clinical signs and symptoms persist.

Statistical relevance of antigen detection tests

There are two examples of statistical relevance. One is the influence of incidence on the practicality of the test, the other, the statistical significance of sensitivity and specificity.

Ryan et al (1986) pointedly demonstrated that the Chlamydiazyme test (Abbott Diagnostics, N. Chicago, IL) for detection of Chlamydia trachomatis, while very useful for certain populations of patients at high risk for sexually transmitted diseases (sensitivity, > 90%; specificity, approaching 99%) is not universally applicable to all populations. Populations, such as asymptomatic male patients, showed test sensitivity of ± 50%. In order for a laboratory test to be applied to a certain population, the test must be evaluated in that population. While such a statement may seem self-evident, there are innumerable instances of available tests being applied to populations representing various risk groups, without having been evaluated in that particular risk group. The commonly used statistic of "predictive value", then, must be interpreted with care. While sensitivity and specificity do not affect predictive value, the prevalence of the disease in the population does. This is precisely why a test cannot be applied equally to all populations. For example, a diagnostic test may have predictive value of 90% in a population where the disease prevalence is 40%. However, when the prevalence of the disease is at 1%, predictive value will decrease precipitously.

The second example is demonstrated by another study, Tilton et al (1988), on the comparison of two direct fluorescent antibody products for C. trachomatis detection. The 2 products were the Kallestad Pathfinder (Kallestad, Chaska, MN) versus MicroTrak (Syva, Palo Alto, CA). The sensitivity of the Syva test is 77% and the Pathfinder, 82%. Without statistical analysis, there is a suggestion that the 2 tests are different, that is, one test more sensitive than the other. However, if the mean and range of sensitivity observed for the 400 patients is calculated and a test for significance, such as the T-test performed, one is surprised to discover that there is no statistically significant difference between the sensitivities of the 2 products. At least 500 patients should be used to determine statistical validity of differences between tests.

Changing patterns of diagnosis and treatment by the use of new modalities for detection of bacteria

The example used to illustrate this comment is the author's experience with the Gen-Probe DNA probe for detection of Mycoplasma pneumoniae in clinical specimens. The laboratory diagnosis of Mycoplasma pneumoniae is often difficult because of lengthy and complicated cultural methods and serological tests that may be both insensitive and non-specific. A study by Tilton et al (1988) on 82 patients suspected of Mycoplasma pneumoniae compared the DNA probe with culture of M. pneumoniae. The probe test was 100% sensitive and 98% specific compared to culture and was completed in 2 hours vs. 14-21 days for the culture of M. pneumoniae. The study indicates that there is essential equivalency between DNA probe and culture of Mycoplasma pneumoniae. It is evident that the timely diagnosis of Mycoplasma primary atypical pneumonia has been hindered by lack of routinely available culture methods for the isolation of the organism from clinical specimens. Factoring in the fact that serologic methods are unreliable leads to the conclusion that the value of rapid diagnosis and early treatment may not have been recognized. However, it has been reported (Jacobs et al, 1986) that treatment with tetracycline or erythromycin significantly reduced the symptoms of mycoplasma pneumonia as well as the length of hospitalization compared to untreated controls. The issue is that as clinicians and laboratorians become more familiar with rapid, precise, and sensitive diagnosis of Mycoplasma pneumoniae and aware that early treatment reduces symptoms, then the test will become a valuable asset in the diagnosis of this widespread disease. This is a rather classic case of the utility of a test such as the DNA probe for the detection of bacteria where the traditional method is complicated and slow. However, the use of this rapid test may not be universally accepted until it becomes evident that it is both clinically useful and cost effective.

In summary, there are a variety of methods for the rapid detection of bacteria, only a few of which have been covered in this essay. However, 3 points must be made: 1) The availability of a test for antigen detection does not guarantee its clinical relevance. 2) More attention must be paid to the statistical relevance of tests for bacterial detection. 3) The advantages of a rapid test compared to a very lengthy traditional test for bacterial detection may not be obvious unless there is an organized method of education and accompanying justification, such as treatment efficacy, for the use of this new test.

REFERENCES

Doern, G.V. 1985, Branhamella catarrhalis: An emerging pathogen. Clin. Micro. Newsl., 7:75-78.

Ellner, P. 1987, Cost containment strategies for the diagnostic microbiology laboratory. Clin. Micro. Newsl., 9:117-120.

Gerber, M.A. 1985, Critical appraisal of the clinical relevance of rapid diagnosis in pediatrics. Diag. Microbiol. Inf. Dis., 3:39S-50S.

Jacobs, E., Bennewitz, A., and Bredt, W. 1986, Reaction pattern of human anti Mycoplasma pneumoniae antibodies in enzyme-linked immunosorbent assays and immunoblotting. J. Clin. Microbiol., 23:517-522.

Ryan, R.W., Kwasnik, I., Steingrimsson, O., Gudmundsson, J., Thorarinsson, H., and Tilton, R.C. 1986. Rapid detection of Chlamydia trachomatis by an enzyme immunoassay method. Diag. Microbiol. Inf. Dis., 5:225-234.

Tilton, R.C., Dias, F., and Ryan, R.W. 1984, Comparative evaluation of three commercial products and counterimmunoelectrophoresis for the detection of antigens in cerebrospinal fluid. J. Clin. Microbiol., 20:231-234.

Tilton, R.C., Barnes, R.C., Gruninger, R.P., Judson, F.N., Ryan, R.W., and Steingrimsson, O. 1988, Multicenter comparative evaluation of two rapid microscopic methods and culture for detection of Chlamydia trachomatis in patient specimens. J. Clin. Microbiol., 26:167-170.

Tilton, R.C., Dias. F., Kidd, H., and Ryan, R.W. 1988, A DNA probe vs. culture for detection of Mycoplasma pneumoniae in clinical specimens. Diag. Microbiol. Inf. Dis., (submitted).

RAPID DIAGNOSIS OF VIRAL INFECTIONS

Thomas F. Smith

Section of Clinical Microbiology
Mayo Clinic and Mayo Foundation
Rochester, MN 55905

INTRODUCTION

Rapid tests for use in the virology laboratory have evolved because of
the recognition of serious complications produced by viruses and the
availability of antiviral therapy such as acyclovir and gancyclovir to
influence the clinical course of the patient. Accordingly, many of these
systemic viral infections occur as a result of impaired host immune defense
mechanisms. Thus, the importance of the herpesviruses in particular, producing life-threatening disease in organ transplant recipients (kidney,
bone marrow, liver, heart, lung, and pancreas) has been recognized for
several years. The severe immunosuppression induced in patients infected
with human immunodeficiency virus (HIV) has amplified the importance of
underlying infection with herpesviruses, and hence laboratory diagnosis of
these infections in this population.

Key technologic developments such as monoclonal antibodies and genetic
probes, many of which are now commercially available, have broadened the
scope of laboratories to detect medically-important viruses with greater
sensitivity and specificity than in years past. Importantly, rapid tests
are now in place for routine diagnosis that may produce results faster than
available in other areas of clinical microbiology.

The extent, direction, and function of diagnostic virology depends on
the type of medical practice that is supported by that laboratory. For
example, rotavirus and many agents causing respiratory tract infections
(respiratory syncytial virus (RSV), influenza virus, parainfluenza virus)
are of much greater concern in primary care medical centers and children's
hospitals than in tertiary care facilities in which the herpesviruses are
likely to predominate. Similarly, practical considerations involving
resources (physical space, laboratory personnel), demography, and numbers
of patients will figure into the equation of the type and extent of laboratory tests performed for virology. Unfortunately, a single rapid technique
is not universally appropriate for the rapid detection of every virus.
This review emphasizes tests primarily used for the laboratory detection of
the herpesviruses.

Laboratory Tests

Cell culture. Recovery of viruses in cells grown in vitro continues to be the standard by which the performance of other tests are measured. Recognition of cytopathic effects (CPE) produced by agents such as herpes simplex virus (HSV), some enteroviruses, and influenza virus can yield diagnostic reports within a 1 to 3 d period. Importantly, these results are highly specific but frequently, in order to achieve the maximum sensitivity, cultures must be incubated and examined for several days.

In general, the disadvantages of cell culture technology include: (i) dependence on efficient transport of specimens to ensure active virus for replication, (ii) preparation of cultures is expensive and labor-intensive, (iii) recognizable CPE may require several days to develop, and (iv) limited susceptibility of host cells to viruses such as rotavirus, Norwalk-like agents, papillomavirus, parvovirus, many coxsackievirus type A strains, hepatitis viruses, and Epstein-Barr virus and many others. Because of these problems, laboratories have sought other more rapid methods that may be used preferably independently, but usually to supplement the conventional tube cell culture technique.

Cytology. For many years, the rapid diagnosis of cytomegalovirus (CMV) was achieved by the detection of characteristic inclusion-bearing cells in stained tissue sections. However, ultimate recovery of the virus in cell cultures, although much slower than the histopathologic technique, has been more sensitive. For example, CMV was recovered from 34 (6.8%) of 502 lung specimens from an unselected autopsy population (Macasaet et al., 1975). Histologic examination revealed the typical intranuclear CMV inclusion bodies in several organs from 12 cases (35%), but in lung tissue from only 6 (18%).

"Similarly, (Moseley et al., (1981)"enrolled 103 consecutive patients presenting to a sexual transmitted disease clinic for suspected genital HSV infection. Specimens from each patient were processed in the laboratory to compare viral isolation, indirect immunoperoxidase, and cytology. HSV was isolated from 81% of these patients and was detected by indirect immunoperoxidase in 57.3%. Papanicolaou smears from the genital lesions revealed giant cells or cells with intranuclear inclusions in 38% of specimens. The maximum sensitivity of detecting a viral infection by cytology depends on the collection of an adequate number of intact cells (25-50) for examination, quality of reagents, and the expertise required for accurate interpretation of results. Conversely, cell culture isolation of the virus is more sensitive, owing to the amplification of virus replication in the host cells over several days period of time.

Electron Microscopy (EM). The technique of negative staining of specimens preparatory to examination of the specimen by the electron microscope was first used to obtain fundamental information about virus morphology. More recently, the procedure has been applied for diagnostic purposes. A drop of a virus suspension is mixed with a 25% aqueous solution of glutaraldehyde on a piece of dental wax. A Formvar-coated, carbon-stabilized grid is floated on the drop for approximately 30 seconds. The grid is then placed on a drop of heavy metal solution, such as phosphotungstic acid. By this procedure, the virus particles are surrounded by heavy metal atoms which act as an electron stain. The electron beam passes through the low density of the virus but not through the metallic background, resulting in the negative stain (Almeida, 1980).

EM has probably been used most extensively for the examination of feces from patients with gastroenteritis. The technique allows recognition of several viruses that are not easily cultivated in cell cultures: Norwalk agents, rotaviruses, adenoviruses, coronaviruses, astroviruses, and caliciviruses (Chernesky et al., 1982). Although EM techniques have been used for the diagnosis of HSV and CMV, most laboratories recognize that the procedure is too expensive, cumbersome, and insensitive for routine viral diagnosis and its utility is limited to special situations (Miller and Lang, 1982; Montplaiser et al., 1972; Richman et al., 1984).

Enzyme Immunoassay (EIA). Conjugation of horseradish peroxidase to immunoglobulin was reported over 20 years ago (Nakane and Pierce, 1966). Enzyme labels are less costly, simpler to use, safer, and have a long reagent shelf life relative to radioisotope tags. In addition, the amplification resulting from the enzyme-mediated conversion of many substrate molecules to a visible chromogenic product by a single enzyme molecule has produced a sensitivity in assay comparable to that of radioimmunoassays (Josephson, 1985). The principle of EIA tests is like that of fluorescent antibody methods, except that the end-point of the assay is a color change that can be objectively measured spectrophotometrically. Similarly to the fluorescent antibody system, both direct and indirect tests can be used to detect viral antigens. The EIA procedure is most often used for detection of viral antigens that are captured by homologous antibodies bound to a solid phase such as a microtiter plate or a bead. The captured antigen is then detected (in the direct test) with another viral-specific antibody conjugated to an enzyme.

EIA's for the direct detection of HSV from clinical specimens have been developed, but generally have lacked the high sensitivity that is needed for the diagnosis of this important virus. For example, an evaluation of the Ortho EIA test (Ortho, Diagnostic Systems, Inc., Raritan, NJ) in our laboratory found a sensitivity of 64% compared to recovery of the virus in conventional tube cell cultures (Morgan and Smith, 1984). Another evaluation of a non-marketed research prototype assay for detection of HSV yielded a sensitivity of 77% (van Ulsen et al., 1987). Amplification of HSV in specimens by replication in cell cultures, however, yielded a much more sensitive system (98%) (Michalski, 1986). Interestingly, another commercial system, "Herp Chek™, DuPont Co., Billerica, MA) was able" to achieve a high detection rate (sensitivity, 95%) of HSV, presumably because of the presence of a viral lysing agent incorporated into the transport medium (Baker et al, 1987).

Nerurkar et al. (1984) modified an EIA test for the direct detection of HSV in clinical specimens by using an enzyme labeled with streptavidin that combined with a biotin-linked anti-HSV antibody that was used as the second antibody (sensitivity, 96%). The biotin-streptavidin system has a very high binding constant (10^{13}) that apparently increases the sensitivity of the EIA. Nevertheless, use of the biotin-avidin EIA in other studies did not necessarily assure the laboratory of highly sensitive and specific results (Adler-Storthz et al., 1983). Variation in test performance may be due to the type of specimen (i.e. titer of virus present) and the relative susceptibility of the assay systems used to detect the virus. In this regard, it is probably much more difficult to develop a highly sensitive EIA test for direct detection of HSV since this virus is very easily recovered in cell cultures. On the other hand, the performance characteristics of similar tests for varicella-zoster virus (VZV) and RSV may be apparently better because the "gold-standard" of cell culture isolation may not be optimal (Smith, 1987).

Immunofluorescence. Cell scrapings from several sources (eye, skin, throat washings), impression smears, and tissue sections may be examined for the presence of viral antigens using fluorescence microscopy. The indirect test, theoretically, is considered to be more sensitive than the direct test since an antispecies antibody reacts with a viral-specific antibody in a sandwich technique. The merit of immunofluorescence tests for antigen detection is the speed of results. The success of the assay, however, depends on the control of many variables: (i) like cytology,

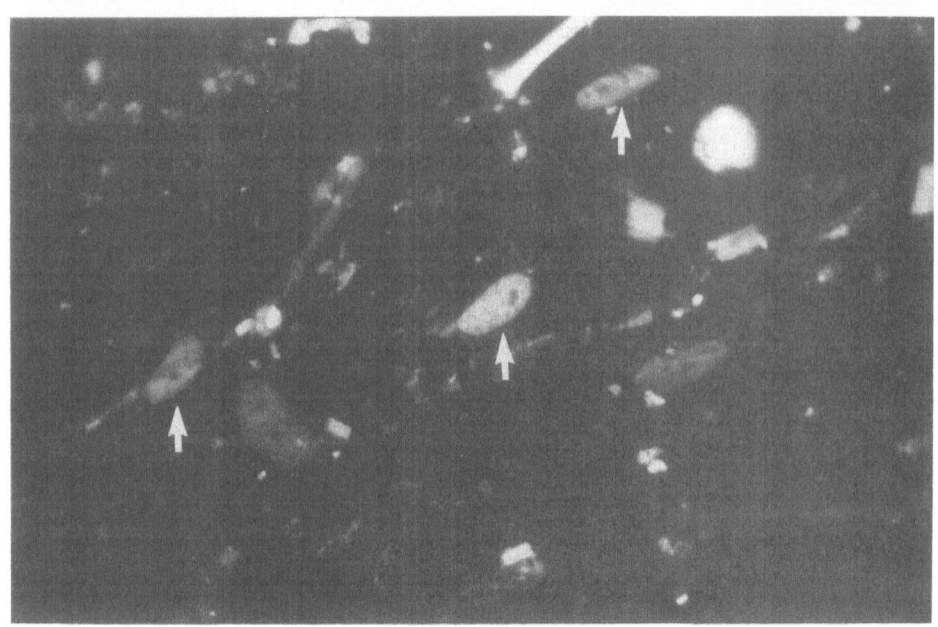

Fig. 1. Foci of CMV-infected MRC-5 cells 16 h postinoculation stained with monoclonal antibody in the indirect immunofluorescence test.

fluorescent antibody tests require an adequate number of intact cells, (ii) quality reagents, (iii) subjective interpretation based on expertise of technologist, and (iv) high-quality fluorescence equipment with optimal filter systems.

Typical experience with the fluorescent antibody test in a pediatric population found that the assay had a sensitivity of 86% for detecting measles, 77% for RSV, 68% for parainfluenza type 3, and 50% or less for adenovirus and influenza virus (Hallsworth and McDonald, 1985). Results were generally less productive when specimens from adults were obtained. Therefore, although the fluorescent antibody test performed on intact cells obtained directly from the patient can be rapid (although tests are performed in batch form in most laboratories), another assay must be performed as backup to provide maximum sensitivity.

Shell Vial Assay. This test is based on the amplification of virus in specimens by short-term infection of cell cultures and the immunologic detection of early antigens by highly specific monoclonal antibodies (Gleaves et al., 1984). The method takes advantage of the immunologic assays (enzyme and fluorescence) to detect viral antigens in their initial stages of production. Early antigens in infected cells are identified 16 h postinoculation by the presence of homogeneous fluorescence throughout the nuclei (Fig. 1). Both the smooth regularity of the nuclear membrane and the characteristic shape of the CMV-infected cell nuclei allow for the specific recognition and distinction of this viral infection from background nonspecific debris that may fluoresce. The availability of a monoclonal antibody with these characteristics for the rapid detection of CMV and the importance of the diagnostic results to clinicians involved in managing organ transplant patients provided impetus to develop a rapid laboratory test for this virus infection.

Many of the steps used in the procedure for isolation of Chlamydia trachomatis have been instructive for achieving maximum sensitivity in the shell vial assay. Several technical aspects of the test must be followed to achieve the desired results and have been discussed in detail in other publications (Gleaves et al., 1984; Paya et al., 1987; Shuster et al., 1985; Smith, 1987). These variables include: age, cell density, and number of cell cultures (shell vials or microtiter plate wells) used per specimen), centrifugation, specificity of monoclonal antibodies for immediate-early or early antigens, type of specimen (urine, blood, bronchoalveolar lavage, tissue), quality of fluorescence equipment, and technical experience with the assay to subjectively evaluate specific results. While adherence to optimal conditions for all these variables is required for optimal performance results, centrifugation of the specimen onto the monolayer is probably the single most important step. Without this enhancement procedure, the sensitivity can be reduced up to 60% (Gleaves et al., 1984). One common problem that many laboratories have experienced during the course of establishing shell vial techniques is the toxicity produced by some specimens. Specimen toxicity of shell vial cell monolayers is likely due to the use of cultures from commercial sources that are already several days old by the time they arrive in the laboratory. For example, Thiele et al. (1987) found that the sensitivity of detection of CMV decreased precipitously with cell cultures that were prepared over 11 days prior to use.

Experience of almost 5 years in our laboratory has indicated that the shell vial assay is more rapid (16 h postinoculation), sensitive, and as specific than conventional tube cell cultures (mean of 9 d for detection) for detection of CMV. Similarly, the technique has been expanded for the diagnosis of HSV, VZV, adenovirus, and influenza virus. These tests are routine procedures in our laboratory in that urine (CMV) specimens and genital specimens (HSV types 1 and 2) are inoculated into 2 shell vials. Dermal specimens are inoculated into 4 shell vials: HSV types 1 and 2 and VZV (stained at 2 d and 5 d postinoculation). Specimens from the eye are inoculated into four shell vials; cells on two coverslips are stained for detection of HSV 1 and 2 the day following inoculation. The remaining two vials are processed for the detection of adenovirus 2 and 5 days postinoculation.

Nucleic Acid Hybridization. Techniques involving nucleic acid hybridization have been known for several years. For example, almost 8 years ago, an assay was described that could accurately differentiate HSV type 1 from type 2 (Brautigan et al., 1980). The principle of the hybridization assay is based on the detection of genetic sequences of viruses present in clinical material with homologous fragments of nucleic acid. These probes are particularly valuable for the detection of many viruses that are not routinely recovered in cell cultures. Their commercial availability will add a new dimension to the diagnostic tests now available in the clinical laboratory.

CONCLUSIONS

Diagnostic virology is in a dynamic state. The incidence of infections in immunocompromised hosts, the rapidly evolving immunologic and nucleic acid probe reagents for diagnosis, and the development of effective antiviral therapy offer limitless opportunity in this area of laboratory practice.

REFERENCES

Adler-Storthz, K., Kendall, C., Kennedy, R. C., Henkel, R. D., and Dreesman, G. R., 1983, Biotin-avidin-amplified enzyme immunoassay for detection of herpes simplex virus antigen in clinical specimens, J Clin Microbiol., 18:1329.

Almeida, J. D., 1980, Practical aspects of diagnostic electron microscopy, Yale J Biol Med., 53:5.

Baker, D. A., Pavan-Langston, D., Gonick, B., Milch, P. O., Dunkel, E. C., Berkowitz, A., Vander-Mallie, R., Woodin, M. B., Philip, A., Bobrow, M. N., and Cukor, G, 1987, Multicenter clinical evaluation of DuPont's Herpchek™, a new rapid diagnostic test for herpes simplex virus (HSV), Abstract, Eastern Pennsylvania Branch, ASM, Philadelphia, PA.

Brautigan, A. R., Richman D. D., and Oxman, M. N., 1980, Rapid typing of herpes simplex virus isolates by deoxyribonucleic acid: deoxyribo-nucleic acid by hybridization, J Clin Microbiol., 12:226.

Chernesky, M. A., Ray, C. G., Smith, T. F., and Drew, W. L., 1982, Laboratory diagnosis of viral infections, Cumitech, 15:1.

Gleaves, C. A., Smith, T. F., Shuster, E. A., and Pearson, G. R., 1984, Rapid detection of cytomegalovirus in MRC-5 cells inoculated with urine specimens by using low-speed centrifugation and monoclonal antibody to an early antigen, J Clin Microbiol., 19:917.

Hallsworth, P. G., and McDonald, P. J., 1985, Rapid diagnosis of viral infections with fluorescent antisera, Pathology, 17:629.

Josephson, S. L., 1985, Immunoenzyme techniques, Clin Microbiol Newsletter, 7:81.

Macasaet, F. F., Holley, K. E., Smith, T. F., and Keys, T. F., 1975, Cyto-megalovirus studies of autopsy tissue. II. Incidence of inclusion bodies and related pathologic data, Am J Clin Pathol., 63:859.

Michalski, F. J., Shaikh, M., Sahraie, F., Desai, S., Verano, L., and Vallabhanine, J., 1986, Enzyme-linked immunosorbent assay spin amplification technique for herpes simplex virus antigen detection, J Clin Microbiol., 24:310.

Miller, S. E., and Lang, D. J., 1982, Rapid diagnosis of herpes simpex infection: amplification for electron-microscopy by short-term in vitro replication, J Infect., 4:37.

Montplaisir, S., Belloncik, S., Leduc, N. P., Angi, P. A., Martineau, B., and Kurstak, E., 1972, Electron microscopy in the rapid diagnosis of cytomegalovirus: ultrastructural observation and comparison of methods of diagnosis, J Infect Dis., 125:533.

Morgan, M. A., and Smith, T. F., 1894, Evaluation of an enzyme-linked immu-nosorbent assay for the detection of herpes simplex virus antigen, J Clin Microbiol., 19:730.

Moseley, R. C., Corey, L., Benjamin, D., Winter, C., and Remington, M. J., 1981, Comparison of viral isolation, direct immunofluorescence and indirect immunoperoxidase techniques for detection of genital herpes simplex virus infection, J Clin Microbiol., 13:913.

Nakane, P. K., and Pierce, G. B., 1966, Enzyme-linked antibodies: prepara-tion and application for localization of antigens, J Histochem Cytochem., 14:929.

Nerurkar, L. S., Namba, M., Brashears, G., Jacob, A. J., Lee, Y. J., and
    Sever, J. L., 1984, Rapid detection of herpes simplex virus in cli-
    nical specimens by use of a capture biotin-streptavidin enzyme-
    linked immunosorbent assay, J Clin Microbiol., 28:109.

Paya, C. V., Wold, A. D., and Smith, T. F., 1987, Detection of cytomegalo-
    virus infections in specimens other than urine by the shell vial
    assay and conventional tube cell cultures, J Clin Microbiol.,
    25:755.

Richman, D. D., Cleveland, P. H., Redfield, D. C., Oxman, M. N., and Wahl,
    G. M., 1984, Rapid viral diagnosis, J Infect Dis., 149:298.

Shuster, E. A., Beneke, J. S., Tegtmeier, G. E., Pearson, G. R., Gleaves,
    C. A., Wold, A. D., and Smith, T. F., 1985, Monoclonal antibody for
    rapid laboratory detection of cytomegalovirus infections: charac-
    terization and diagnostic application, Mayo Clin Proc., 60:577.

Smith, T. F., 1987, Rapid methods for the diagnosis of viral infections,
    Lab Med., 18:16.

Thiele, G. M., Bicak, M. S., Young , A., Kinsey, J., White, R. J., and
    Purtilo, D. T., 1987, Rapid detection of cytomegalovirus by tissue
    culture, centrifugation, and immunofluorescence with a monoclonal
    antibody to an early nuclear antigen, J Virol Meth., 16:327.

van Ulsen, J., Dumas, A. M., Vagenvoort, J. H. T., van Zuuren, A., van
    Joost, T., and Stolz, E., 1987, Evaluation of an enzyme immunoassay
    for detection of herpes simplex virus antigen in genital lesions,
    Eur J Clin Microbiol., 6:410.

TIME-RESOLVED FLUOROMETRY: PRINCIPLES AND APPLICATION TO CLINICAL

MICROBIOLOGY AND DNA PROBE TECHNOLOGY

Pertti Hurskainen, Patrik Dahlén, Harri Siitari
and Timo Lövgren

Wallac Oy
P.O. Box 10
SF - 20101 TURKU
Finland

INTRODUCTION

Radioactive isotopes are the most widely used markers in clinical che-
mistry. Although radioactive labels are very sensitive they have several
drawbacks. The most severe disadvantages are the health hazards involved
and the limited shelf life of reagents labeled with radioactive markers.
In order to overcome these difficulties, alternative nonisotopic labels
have been developed. Among the nonradioactive markers are enzymes, lumi-
nescent compounds and fluorescent probes.

Theoretically highly sensitive fluorescent labels have not gained wide
use. The problem has been the inferior sensitivity which is mainly due to
the high background in fluorometric measurement. Combination of rare earth
metals and the time-resolved measurement principle has lowered the background to a level which makes very sensitive assays possible[1-3].

The most important factors in the detection of microbes are sensitivity and specificity. Immunological methods in the diagnosis of microbial
infections are well established. Highly sensitive time-resolved fluoroimmunoassay (TR-FIA) principle has been applied to the detection of viral
antigens[4-7] and antiviral antibodies[8,9].

DNA hybridization is an important new technique in the diagnosis of
infectious diseases. However, the use of radioisotopes as labels and the
laborious multi-step procedure have prevented the transfer of DNA probe
analysis to routine clinical laboratories. A number of attempts to replace
radioisotopes have been made. The most commonly employed nonradioactive
marker is biotin. Biotin on the probe can be detected indirectly using
labeled avidin or streptavidin[10-12]. However, biotin as a marker has not
replaced radioisotopes because biotinylated probes are not as sensitive as
^{32}P-labeled probes[13,14] and may give false-positive results with microbial samples[14].

Time-resolved fluorometry with lanthanide chelates was reviewed recently[15]. We briefly describe the principles of time-resolved fluorometry
and focus on the application of this new technique to clinical
microbiology and DNA probe technology.

Rapid Methods in Clinical Microbiology
Edited by B. Kleger *et al.*
Plenum Press, New York

The fluorescence properties of lanthanides make them potentially very useful as labels. Lanthanides have a broad excitation and narrow emission band. The difference between excitation and emission maxima (Stokes shift) of Eu^{3+} and Tb^{3+} is more than 200 nm.[15] The fluorescence decay time of rare earth metals, especially that of Eu^{3+} and Tb^{3+} is long, ranging from 1 μs to 1 ms (Fig. 1).[15]

Background in fluorometric assays consists mainly of fluorescence and scattering. Background fluorescence caused by impurities and fluorescent compounds in reagents and cuvettes has typically a short decay time. The excitation light can be scattered from soluble molecules, small particles or the solid-phase material. The Stokes shift of scattering is not more than 50 nm.[3]

The properties of lanthanides make it possible to distinguish a specific signal from the background noise. Although the quantum yield of the best organic fluorescent compounds is at least 10-fold as compared to lanthanide chelates the overall sensitivity of rare earth metal chelates is comparable or better.[15]

The fluorescence of inorganic salts of lanthanides is weak. When the lanthanide ion (M^{3+}) is bonded to suitable organic ligands, fluorescence can be dramatically enhanced. The best organic ligands in this respect are the β-diketones. However, complexes of lanthanides with β-diketones are not stable enough to be used as labeling reagents. In the DELFIA[TM] system (Wallac Oy, Finland) the reagents are labeled with stable nonfluorescent lanthanide chelates. After completing the assay, the lanthanide is dissociated and measured as a highly fluorescent β-diketone chelate in the presence of synergistic additives. In the case of europium, the excitation wavelength is 340 nm and the duration of excitation pulse is 1 μs. After a delay time of 400 μs fluorescence is counted for 400 μs at 613 nm using a single-photon counting fluorometer. The cycle is repeated 1000 times during the total counting time of 1 s (Fig. 2).[1],[15]

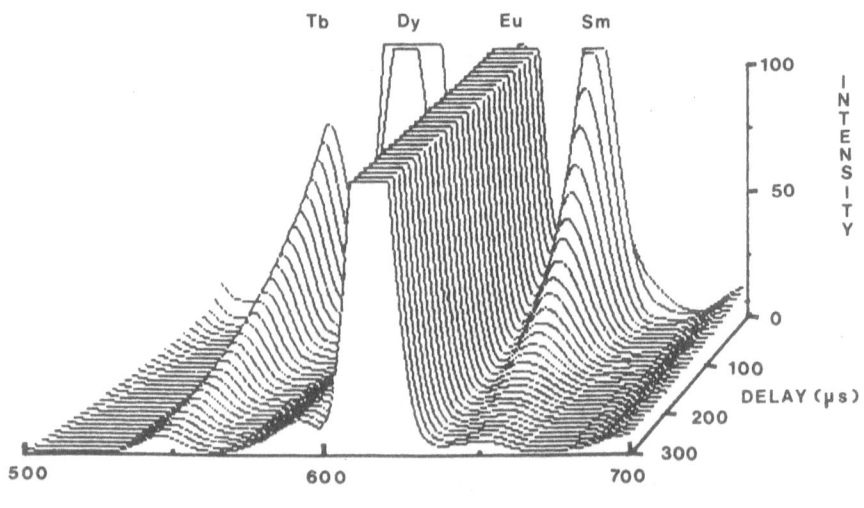

EMISSION (nm)

Figure 1. Fluorescence decay of europium (Eu), terbium (Tb),
samarium (Sm) and dysprosium (Dy).

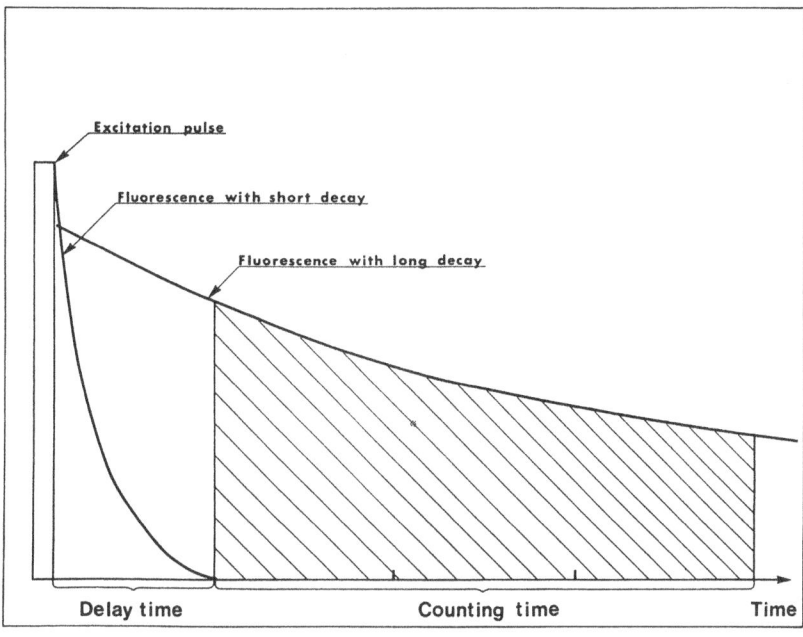

Figure 2. Operation principle of time-resolved fluorometer.

APPLICATION OF TIME-RESOLVED FLUORESCENCE IN VIRAL DIAGNOSIS

Very sensitive solid phase immunoassays have been developed for the testing of viral antigens in clinical specimens. Radioactive tracers[16-19] and enzyme labels[19-22] have been widely used in these assays. Previously, the indirect detection principle was often applied in virus antigen detection.[5,19,20] This principle involves laborious assay procedures with e.g. three incubations and an unnecessarily long delay in obtaining the results.

Virus Tests Based on Eu[3+]-Labeled Antibodies

We applied the high specific activity of the Eu[3+] chelate in developing an assay in which high sensitivity is needed. Hepatitis B surface antigen (HBsAg) was chosen for a model where it is clearly beneficial to detect the antigen as early as possible at an early phase of the disease. In addition, the potentially infectious phase of the disease can be followed longer into convalescence. This shortens the window period where the surface antigen markers are usually not detectable. HBsAg screening is also done in large scale in blood banks, blood fractionating centers etc. where the test should be as simple as possible.[4] The Delfia[TM] HBsAg assay has a sensitivity of at least 0.2 ng/ml and 0.05 ng/ml with one incubation and two incubation procedures respectively. Both ad and ay subtypes can be detected effectively by using monoclonal Eu[3+]-labeled antibodies which are specific for the main determinant a, and polyclonal antibodies derived from sheep in the solid phase. A sensitivity panel from OoB (Office of Biologics of the Food and Drug Administration) has also been tested (Table 1). All the positive specimens from A to D were positive.

Time-resolved fluorescence has been compared to radioimmunoassays in the detection of adenovirus and rotavirus in stool specimens and respiratory syncytial virus, Parainfluenza-1 and adenovirus in nasopharyngeal aspirates[5]. The indirect assay principle and polystyrene beads as the

125

Table 1. OoB 500 panel was tested with two-incubation procedure.
All specimens were run in duplicate and positive control
in three duplicates, negative control with sera duplicates

Specimen code	Cps	Category
501 B	1728 770	+
502(C)	26 149	+
503 C	184 673	+
504 C	422 813	+
505 C	68 832	+
506 N	1 443	−
507 C	61 098	+
508(C)	4 719	+
509 D	2 383	+
510 N	1 486	−
511 C	39 460	+
512 D	2 970	+
513 C	538 662	+
Pos. control	32 167	+
Neg. control	1 375	−
Cut-off value*	1 925	

*cut-off value is defined as 1.4 times the
mean of negative controls according to the kit
insert.

solid phase were used in these assays. The assay sensitivities were equal with TR-FIA and RIA.

The one-incubation assay procedure was then also adapted to respiratory virus detection in nasopharyngeal aspirates and adeno- and rotavirus detection in stool specimens[6],[7]. Microtitration strips (12 wells/strip) were used as solid phase in order to further simplify the washings and overall handling during the test. This assay format was used in comparison to ELISA (Enzyme Linked Immunosorbent Assay) and TR-FIA in the detection of rota- and adenovirus in stool specimens and adenovirus in nasopharyngeal aspirates.[23] TR-FIA showed higher specificity than ELISA (3 false positives in 50 ELISA positive specimens). The mild conjugation reaction and the very low molecular weight Eu^{3+} chelate[2] does not easily affect the performance characteristics of the antibodies which minimizes possible unspecific side reactions. A very high assay sensitivity was also obtained recently in the detection of adenoviruses[24] and influenza viruses[25] with monoclonal antibodies.

Influenza A virus detection with monoclonal Eu^{3+}-labeled antibodies has been used for cell culture confirmation.[26] This one-incubation one-hour test from nasal washes correlated well with the hemadsorption method when tested from primary monkey kidney cell cultures.

TR-FIA in Viral Serology

The first application was a test for Rubella antibodies.[8] Sensitivity was found to be at least equal to ELISA and RIA tests. In addition, the large assay range together with good linearity allowed the tests to be done with single dilutions. Moreover, antibody concentrations can be easily quantified allowing e.g. the follow-up of acute infection. The

Eu^{3+}-labeled swine antihuman IgG conjugate was used very recently for the determination of antibodies against HIV.[9] In the study the rate of false positives was extremely low, much lower than with ELISA. In fact, no grey zone could be seen, which is also our experience and which allows a clear distinction between positive and negative specimens. This has of course also practical benefits in terms of reducing the number of confirmatory tests.

TIME-RESOLVED FLUOROMETRY IN DNA PROBE TECHNOLOGY

The application of time-resolved fluorometry to detect DNA sequences has been reported. Eu^{3+}-labeled streptavidin has been used with biotinylated probes.[12,27] A hapten-modified probe can be detected with high sensitivity by using a Eu^{3+}-labeled antibody.[28] To simplify the nucleic acid hybridization procedure we developed a method of labeling DNA with a europium chelate. The properties of Eu^{3+}-labeled DNA probes are described. Finally we discuss the sensitivities with different markers.

Labeling of DNA with Europium Chelate

Cytosine residues in single-stranded DNA can be transaminated in the presence of sodium bisulfite and ethylenediamine.[29] The modification yields DNA which contains primary aliphatic aminogroups. Subsequently, transaminated DNA is reacted with the isothiocyanate analogue of the europium chelate W2014. By controlling the transamination reaction it is possible to modify 1-10% of all nucleotides in DNA with Eu^{3+}.

Properties of Eu^{3+}-Labeled DNA

The modification of DNA with chemical groups results in alterations in the properties of DNA. In double-stranded DNA the aminogroups of deoxycytidine residues are involved in hydrogen bonding responsible for specific base pairing. Modification of deoxycytidine residues with a europium chelate decreases the thermal stability of DNA duplexes. Melting temperatures of lambda phage DNA with 5% and 10% Eu^{3+} incorporation were 3.5°C and 7°C, respectively, lower than T_m of unmodified lambda DNA.

Hybridization kinetics and efficiency with Eu^{3+}-labeled lambda phage probes were studied in a mixed-phase system. Target lambda DNA was spotted onto a nitrocellulose filter and hybridized with Eu^{3+}-labeled lambda DNA. Hybridization efficiency of lambda DNA probes with 5% and 10% Eu^{3+} modification was 11% and 8%, respectively. Maximum signal for both probes was achieved in a similar period of time.

Enzymatic biotinylation of DNA with biotin-dUTP was reported to decrease the melting temperature by 0.4°C per each percent of biotin incorporation.[30] Hybridization rate in solution became slower after chemical biotinylation involving transamination.[31] However, chemically labeled probes hybridized efficiently in a mixed-phase system. Similar results were obtained by introducing biotin-dCTP enzymatically into DNA.[32]

Probes with high Eu^{3+} incorporation (~10%) hybridize less efficiently and show a decreased melting temperature but these effects have not been found to impair the specificity of the hybridization reaction.

DNA probes labeled with ^{32}P or ^{125}I have a limited shelf life. Eu^{3+}-labeled probes are stable at -20°C for at least one year.

Sensitivity of Eu^{3+}-Labeled DNA Probes

Eu^{3+}-labeled lambda DNA probe could detect 10 pg (0.3 attomoles) of lambda DNA fixed on either nitrocellulose or polystyrene (Fig. 3). Sen-

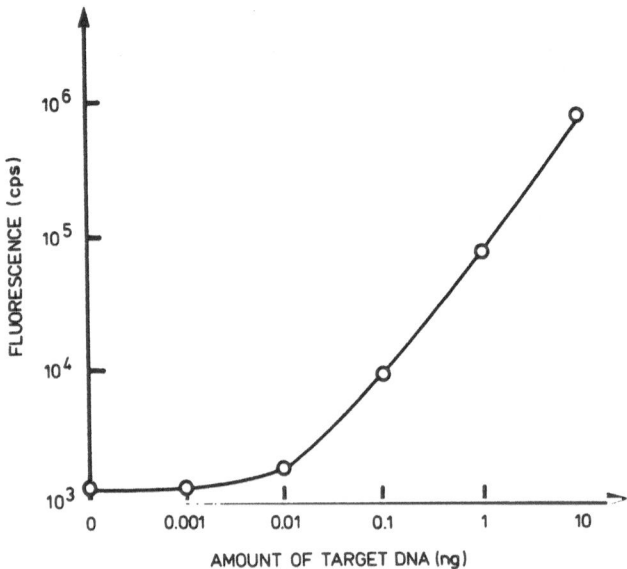

Figure 3. Sensitivity of Eu^{3+}-labeled DNA probes. Eu^{3+}-labeled
lambda DNA (5% incorporation) was hybridized to different
amounts of lambda DNA coated onto polystyrene strips.
Probe concentration 400 ng/ml.

sitivity was the same for probes with 5% and 10% Eu^{3+} modification. The
high sensitivity of Eu^{3+}-labeled probes is partly due to the high probe
concentrations (up to 500 ng/ml) which can be used. High probe concentrations have a beneficial effect on hybridization kinetics and efficiency.

For the sake of comparison we performed hybridizations with a radioactive and a biotinylated lambda DNA probe. A ^{32}P-tagged probe was able to
detect 1 pg of lambda DNA after overnight autoradiography and 10 pg of
lambda DNA after counting in a liquid scintillation counter. A biotin
system using a commercial kit for labeling of the probe and enzymatic
hybrid detection was able to detect 1 ng of lambda DNA.

It has been reported that sensitivity in DNA hybridization with ^{32}P-
labeled probes is 5-100 times better than by using biotinylated
probes.[13],[14] According to our experience the sensitivity of ^{32}P-tagged
probes after overnight autoradiography is repeatedly 100-1000 -fold compared to that of biotinylated probes.

CONCLUSIONS

Time-resolved fluorometry has been found to be a very sensitive method
in the detection of viral antigens and antiviral antibodies. The specificity of TR-FIA is excellent and there is a clear separation between positive and negative specimens Thus a markedly low number of specimens need
to be retested. An additional advantage of TR-FIA is the wide linear measurement range enabling usually testing of only one sample dilution.

In nucleic acid hybridization the lack of sensitive, nonradioactive
markers is a major obstacle to a wide application of DNA-based assays in
routine laboratories. To overcome this problem we developed a method to
label DNA with a europium chelate. Eu^{3+}-labeled probes are as sensitive as
^{32}P-labeled probes detected by liquid scintillation counting. The detection step after hybridization with Eu^{3+}-labeled probes takes only about 20

minutes whereas biotinylated probes require a several hours long enzymatic reaction to obtain the optimal result. Further improvements are, however, needed to simplify hybridization assays. Tests including fixation of samples on solid matrices are inconvenient and susceptible to background problems. Difficulties in sample treatment can be avoided in solution hybridization assays. Solution hybridization also results in a rapid assay because the rate of annealing in solution is much higher than in a mixed-phase system. In our opinion a method based on the combination of solution hybridization and time-resolved fluorometry will be a strong candidate for routine tests in clinical microbiology.

LITERATURE CITED

1. Soini, E. and I. Hemmilä, Fluoroimmunoassay: present status and key problems, Clin. Chem. 25:353-361 (1979).
2. Hemmilä, I., Dakubu, S., Mukkala, V.-M., Siitari, H. and T. Lövgren, Europium as a label in time-resolved immunofluorometric assays, Anal. Biochem. 137:335-343 (1984).
3. Hemmilä, I., Fluoroimmunoassays and immunofluorometric assays, Clin. Chem. 31:359-370 (1985).
4. Siitari, H., Hemmilä, I., Soini, E., Lövgren, T. and V. Koistinen, Detection of hepatitis B surface antigen using time-resolved fluoroimmunoassay, Nature (London) 301:258-260 (1983).
5. Halonen, P., Meurman, O., Lövgren, T., Hemmilä, I. and E. Soini, Detection of viral antigens by time-resolved fluoroimmunoassay, in: New Developments in Diagnostic Virology, P.A. Bachmann, ed., Springer-Verlag, Berlin, (1983).
6. Halonen, P., Bonfanti, C., Lövgren, T., Hemmilä, I. and E. Soini, Detection of viral antigens by time-resolved fluoroimmunoassay, in: Rapid Methods and Automation in Microbiology and Immunology, K.-O. Habermehl, ed., Springer-Verlag, Berlin, (1984).
7. Halonen, P., Bonfanti, C., Waris, M., Lövgren, T. and I. Hemmilä, New developments in diagnostic virology, in: Proc. European Symp. New Horizons in Microbiology, A. Sanna and G. Morace, ed, Elsevier/North Holland, Amsterdam, (1984).
8. Meurman, O.H., Hemmilä, I.A., Lövgren, T.N. and P.E. Halonen, Time-resolved fluoroimmunoassay: A new test for rubella antibodies, J. Clin. Microbiol. 16:920 (1982).
9. Aceti, A., Titti, F., Verani, P., Butto, S., Pennica, A., Sebastiani, A. and G.B. Rossi, Time-resolved immunofluorescence: a sensitive and specific assay for anti-HIV antibody detection in human sera, J. Virol. Methods 16:303-315 (1987).
10. Leary, J.J., Brigati, D.J. and D.C. Ward, Rapid and sensitive colorimetric method for visualizing biotin-labeled probes hybridized to DNA or RNA immobilized on nitrocellulose: Bio-blots, Proc. Natl. Acad. Sci. USA 80:4045-4049 (1983).
11. Matthews, J.A., Batki, A., Hynds, C. and L.J. Kricka, Chemiluminescent method for the detection of DNA in dot-hybridization assays, Anal. Biochem. 151:205-209 (1985).
12. Dahlén, P., Detection of biotinylated DNA probes by using Eu-labeled streptavidin and time-resolved fluorometry, Anal. Biochem. 164:78-83 (1987).
13. Hyypiä, T., Detection of adenovirus in nasopharyngeal specimens by radioactive and nonradioactive DNA probes, J. Clin. Microbiol. 21:730-733 (1985).
14. Zwadyk, P., Cooksey, R.C. and C. Thornsberry, Commercial detection methods for biotinylated gene probes: Comparison with ^{32}P-labeled DNA probes, Curr. Microbiol. 14:95-100 (1986).
15. Soini, E. and T. Lövgren, Time-resolved fluorescence of lanthanide probes and applications in biotechnology, CRC Crit. Rev. Anal. Chem. 18:105-154 (1987).

16. Ling, C.M. and L. E. Overby, Prevalence of hepatitis B virus antigen revealed by direct radioimmunoassay with ^{125}I-antibody, J. Immunol. 109:834-841 (1972).

17. Cameron, C.H., Combridge, D.R., Howell, D.R. and J.A.J. Barbara, A sensitive immunoradiometric assay for the detection of hepatitis B surface antigen, J. Virol. Meth. 1:311-323 (1980).

18. Ben-Porath, E., Wands, J., Gruia, M. and K. Isselbacher, Clinical significance of enhanced detection of HBsAg by a monoclonal radio-immunoassay, Hepatology 4 no: 5:803-807 (1984).

19. Sarkkinen, H.K., Tuokko, H. and P.E. Halonen, Comparison of enzyme immunoassay and radioimmunoassay for detection of human rotaviruses and adenoviruses from stool specimens, J. Virol. Methods 1:331-341 (1980).

20. Sarkkinen, H.K., Halonen, P.E., Arstila, P.P. and A.A. Salmi, Detection of respiratory syncytial, parainfluenza type 2, and adenovirus antigens by radioimmunoassay and enzyme immunoassay on nasopharyngeal specimens from children with acute respiratory disease, J. Clin. Microbiol. 13:258-265 (1981).

21. Chao, R.K., Fishaut, M., Schwartzman, J.D. and K. McIntosh, Detection of respiratory syncytial virus in nasal secretions from infants by enzyme-linked immunosorbent assay, J. Inf. Diseases. 139:483-486 (1979).

22. Yolken, R.H., Enzyme-linked immunosorbent assay (ELISA). A practical tool for rapid diagnosis of viruses and other infectious agents, Yale J. Biol. Med. 53:85-92 (1980).

23. Siitari, H., Lövgren, T. and P. Halonen, Detection of viral antigens by direct one-incubation time-resolved fluoroimmunoassay, in: Developments in Applied Biology: Developments and Applications in Virus Testing, vol. 1, R.A.C. Jones and L. Torrace, eds, Association of Applied Biologists, Wellesbourne, (1986).

24. Hierholtzer, J.C., Johansson, K.H., Andersson, L.J., Tsou, C.J. and P.E. Halonen, Comparison of monoclonal time-resolved fluoroimmunoassay with monoclonal capture-biotinylated detector enzyme immunoassay for adenovirus antigen detection, J. Clin. Microbiol. 25:1662-1667 (1987).

25. Walls, H.H., Johansson, K.H., Harmon, M.W., Halonen, P.E. and A.P. Kendal, Time-resolved fluoroimmunoassay with monoclonal antibodies for rapid diagnosis of influenza infections, J. Clin. Microbiol. 24:907-912 (1986).

26. Greene, W.H., Betts, R.F., Siitari, H. and M.A. Menegus, Manuscript in preparation. (1987).

27. Dahlén, P., Syvänen, A.-C., Hurskainen, P., Kwiatkowski, M., Sund, C., Ylikoski, J., Söderlund, H. and T. Lövgren, Sensitive detection of genes by sandwich hybridization and time-resolved fluorometry, Mol. Cell. Probes 1:159-168 (1987).

28. Syvänen, A.-C., Tchen, P., Ranki, M. and H. Söderlund, Time-resolved fluorometry: a sensitive method to quantify DNA-hybrids, Nucleic Acids Res. 14:1017-1028 (1986).

29. Shapiro, R. and J.M. Weisgras, Bisulfite catalyzed transamination of cytosine and cytidine, Biochem. Biophys. Res. Commun. 40:839-843 (1970).

30. Langer, P., Waldrop, A. and D. Ward, Enzymatic synthesis of biotin-labeled polynucleotides: Novel nucleic acid affinity probes, Proc. Natl. Acad. Sci. USA 78:6633-6637 (1981).

31. Viscidi, R.P., Connelly, C.J. and R.H. Yolken, Novel chemical method for the preparation of nucleic acids for nonisotopic hybridization, J. Clin. Microbiol. 23:311-317 (1986).

32. Gillam, J.C. and G.M. Tener, N^4-(6-aminohexyl)cytidine and -deoxy-cytidine nucleotides can be used to label DNA, Anal. Biochem. 157:199-207 (1986).

THE CLINICAL IMPACT OF AUTOMATED SUSCEPTIBILITY

REPORTING USING A COMPUTER INTERFACE

Alan T. Evangelista

Division of Microbiology and Immunology
Cooper Hospital/University Medical Center
Camden, New Jersey   08103

INTRODUCTION

Mechanisms to improve patient care are presently evolving in the clinical laboratory in the areas of rapid testing and computerized reporting of laboratory results. The clinical impact of computerized reports and specifically antimicrobial susceptibility results can also be viewed in terms of their economic impact which is a primary concern for hospitals today with the prospective reimbursement issues of diagnostic related groups (DRG's). The financial restrictions have generated administrative pressures to provide efficient patient care while limiting the use of hospital resources. Thus, mechanisms that allow a physician to make an earlier clinical decision regarding patient care will contribute to reducing the patient's length-of-stay (LOS) thereby benefiting the patient as well as the hospital. The role of the laboratory towards reducing LOS has traditionally been to provide accurate results with as rapid a turnaround time (TAT) as possible.

With the advent of computerization, TAT can now be more accurately measured. Furthermore, the clinical and economic impact of rapid reporting can be measured in both a direct and indirect manner. The direct measurement of a rapid result is achieved by recording the start and stop times of patient antibiotic orders. The timing of antibiotic orders reflects both the microbiology lab TAT for bacterial identification and susceptibility results as well as physician response time to those results. Modifications of antimicrobial therapy have a direct effect on the pharmacy budget and early modifications to narrow spectrum lower cost antibiotics represent economic benefits to the hospital. Moreover, if a patient is placed on a 7 day course of appropriate therapy 1 or 2 days earlier, then the length of stay can be expected to be reduced by 1 to 2 days. However, the impact of a rapid lab result on patient LOS is necessarily an indirect measurement because the decision to discharge a patient on a given day and time is based on physician judgement and is multifactorial. Even so, rapid lab results indirectly contribute to improved utilization of hospital beds and personnel, decreased use of hospital resources outside of the lab and pharmacy, and reduced potential for nosocomial infections. Monitoring the TAT of lab results is also an important parameter for hospital quality assurance programs especially when a susceptibility result can be followed up to determine appropriateness of antibiotic usage.

MICROBIOLOGY RESULT REPORTING

A major transition in any laboratory is the shift from manual result reporting on a multi-copy lab slip to computerized result reporting. Assuming that the laboratory computer is microbiology friendly the reporting efficiency, accuracy and TAT are markedly improved with fewer transcription errors. Additionally, both preliminary and final results can be made available to the medical staff instantly via remote terminals throughout the hospital. An enhancement above technologist keyboard entry of results is the on-line interface from the automated bacterial identification/susceptibility system to the laboratory host computer. Interfacing itself is subject to degrees of efficiency. At one level of operation the microbiology data is released to the nursing station following review by a technologist. However, further software enhancements are available that incorporate user-defined acceptable criteria for identification/susceptibility results and thus allow autofiling of preliminary results without review by a technologist. In the latter case technologist review is necessary to confer a final status on results.

The autofiling of preliminary results is a software option available to all laboratories having an automated Vitek Systems instrument for bacterial identification and susceptibility testing and a Sunquest Information Systems laboratory computer. The Vitek-Sunquest interface and has been used effectively at Cooper Hospital for the past 3 years. Autofiling occurs for an organism/ susceptibility result when the identification confidence level is greater than 85% (most are above 95%). Any drug-bug combination that is perceived as a problem result can be defined in a Sunquest Quality Assurance Report and is automatically removed from the autofiling function. All finalized results are reviewed by the supervisor or director via the Sunquest Culture Review Report, a summary of positive culture results.

AUTOFILING AND TURNAROUND TIME

The Vitek-Sunquest on-line interface autofiling function represents the current state-of-the-art in computerized rapid reporting of microbiology results. Turnaround time can optimally be viewed as the reporting of bacterial identifications and MIC susceptibilities on the day after receipt of specimen. For example, a specimen received on a Monday morning is inoculated onto the appropriate culture media and incubated overnight. On Tuesday morning the culture plates are read, bacterial suspensions are made from isolated colonies and Vitek identification and susceptibility cards are inoculated and placed in the reader/incubator. The majority of results are completed by the automated system between 3 and 5 o'clock in the afternoon, are sent across the on-line interface, and if they meet user-defined criteria are viewable as preliminary results by medical and house staff on remote Sunquest terminals. Emphasis is placed on the 5:00 PM cutoff, because the primary residents in charge of patient care complete their shift around that time. Any modification of antimicrobial therapy is more likely to occur if an attending physician or resident in charge can view results prior to 5:00 PM. To accomodate this cut-off time the Microbiology Dept. operates with two staff schedules for the day shift. The early shift follows a 7:00 AM to 3:30 PM schedule and the later shift 8:30 AM to 5:00 PM The early shift picks isolates from the primary overnight culture plates and sets up Vitek cards so that reportable results are available by midafternoon.

AUDIT OBJECTIVES AND TOOLS

Determining the clinical impact of computerized rapid result reporting involved three objectives each focusing on separate time issues. The first objective was to determine the incubation/processing times required for completion of Vitek test results. The second objective was to determine the time during the day when Vitek results were uploaded across the interface to the Sunquest host computer and specifically the percentage of completed results prior to 5:00 PM. The third objective was to investigate the timing of anti-biotic orders and to determine if rapid results were used by physicians to modify antimicrobial therapy in a cost effective manner.

The tools or sources of data for this study were gathered from the Vitek, Sunquest and hospital computer systems. Individual Vitek card printouts were used to record the times for completed test results. In addition, Vitek tray directories were printed at 5:00 PM each day during the audit period to determine percentage of completed results. From the Sunquest system were printed Culture Review Reports which summarized all positive culture results, specimen collection times and the dates/times of preliminary and final reports. The antibiotic ordering data was retrieved from the medication order log of the hospital information system which included antibiotic start and stop times, route and dosage information and number of antibiotic units dispensed.

AUDIT OF VITEK COMPLETED RESULTS

The card processing times required for Vitek results to be completed were recorded over a one month period. The results as shown in Table 1 were divided into gram negative identification (GNI) cards and gram nega-tive susceptibility (GNS) cards for the seven most common organisms. The category GNS represented the custom card GNS-J for gram negatives and GNS-P for Pseudomonas. A result was considered completed when both the GNI and GNS test results had been uploaded to Sunquest, therefore the card requiring the longer incubation time was the limiting factor. The organisms listed in Table 1 were divided into two groups. Those that required less than 6 hours for completion were the lactose fermentors with Klebsiella pneumoniae showing the earliest completion time at an average 5.3 hours. The organisms that required greater than 6 hours were nonlactose fermentors and Pseudomonas aeruginosa produced the longest average completion time of 8.1 hours. When Staphylococci were tested using gram positive susceptibility (GPS-M) cards the time to completion was always 6.0 hours.

Table 1. Time to Completed Vitek Result

ORG < 6 HR	GNI	GNS
K. pneumoniae	4.6	5.3
Enterobacter sp.	4.5	5.4
E. coli	4.4	6.0
ORG > 6 HR	GNI	GNS
M. morganii	5.5	6.7
S. marcescens	7.0	6.5
P. mirabilis	7.4	5.6
P. aeruginosa	7.0	8.1

During the two month audit period the percentage of completed Vitek results by 5:00 PM was compiled from Vitek tray directories printed on 24 randomly selected days. A review of a total of 430 card pairs (GNI/GNS) showed that the 5:00 PM completion rate was highly dependent on workload volume. On days of high workload, when 2 to 2 ½ trays of cards were filled, the 5:00 PM completion rate averaged only 40%. Moderate workload days of 1½ to 2 trays resulted in 5:00 PM completion rates averaging 60%, while light workload days of 1 to 1 ½ trays averaged 85% completed by 5:00 PM. With the daily workload usually leaning towards moderate to heavy the overall completion rate for the audit was 55% by 5:00 PM. However, the remainder of the card results were completed by 10:00 PM and were therefore included on the printed Sunquest individual cumulative reports which were placed on the patient's chart very early the following morning.

AUDIT OF POSITIVE BLOOD CULTURES

In determining the impact of rapid computerized results on antimicrobial therapy the initial audit focused on positive blood cultures. Clinical decisions regarding antimicrobial therapy were placed into six categories based on the start and stop times of antibiotic orders. The first category was termed Adequate Empiric Therapy and identified those patients who were placed on antibiotics prior to or shortly after blood cultures were drawn. This study determined only if the empiric therapy would adequately cover the eventual positive blood culture and not whether the therapy was the most cost effective selection. The following four categories identified patients whose antibiotic therapy was modified in response to a reported result and were listed as: Gram Stain, Interface Autofiling, Printed Report and Printed Report + 1 Day. These first five categories were not mutually exclusive. For example, a patient could be placed on adequate empiric therapy and would also be recorded as positive if therapy was modified based on a gram stain result. The same patient would again be recorded as positive if therapy was modified further based on a preliminary result of the Vitek-Sunquest interface autofiling. The audit was thus recording physician responses to information provided by the microbiology lab over time. The sixth and final category was called Inappropriate Therapy which included patients whose antimicrobials had not been modified to cover the positive culture results even after printed results had been charted for 2 days.

A total of 36 patients with documented positive blood cultures were audited at random over a one month period using the computerized Sunquest summary reports and the medication orders recorded in the hospital information system. The results shown in Table 2 indicated that a majority of the patients (56%) with documented bacteremia had been placed on adequate empiric therapy. Modifications in therapy were recorded in 8 patients (22%) following the report of a gram stain result, and in 10 patients (28%) in response to preliminary identification/susceptibility results which had autofiled and were available through Sunquest terminals at nursing stations. Additional antibiotic modifications were made in 4

Table 2. Blood Culture Therapy Audit

A.	Adequate Empiric Therapy	20/36 = 56%
B.	Modification Based on:	
	Gram Stain	8/36 = 22%
	Interface Autofiling	10/36 = 28%
	Printed Report	4/36 = 11%
	Report + 1 Day	2/36 = 6%
C.	Inappropriate Therapy	2/36 = 6%

Table 3. Patient A

2/18/88		Adm. date 71 F recurrent sepsis
2/20	0630	blood culture collected
	0700	start ceftazidime IV 2 g q 8h $78/day
		nafcillin IV 2 g q 4h $29/day
		tobramycin IV 90mg q 8h $33/day
2/21	1100	gram stain result (gnr) called to floor
2/22	1500	autofiled Vitek GNI/GNS K. pneumoniae
	1700	stop ceftaz. naf & tobra $140/day
		start cefazolin IV 1 g 1 8h $17/day

savings $140-17 = $123/day x 10 days = $1,230

patients (11%) following the charting of a printed report and in 2 patients (6%) after the result had been charted for 1 day. Two patients (6%) fell into the category of inappropriate therapy.

An example of the economic impact of autofiling is shown in Table 3. Patient A a 71 year old female was admitted on 2/18/88 with a diagnosis of recurrent sepsis. A blood culture was drawn at 6:30 AM on 2/20/88 and shortly thereafter she was started empirically on triple therapy which included ceftazidime, nafcillin and tobramycin for coverage of gram neg rods, Pseudomonas and Staph. The hospital cost for these antibiotics which included pharmacy and nursing preparation costs totalled $140 per day. On the following morning (day 2) the blood culture bottles were detected as positive, the gram stain result of gram negative rods was called to the nursing station and entered into the computer, and the bottles were subcultured. Early on 2/22/88 (day 3) isolated colonies were obtained and the GNI/GNS Vitek card pair was set up. The cards finalized at 3:00 PM on day 3 and a preliminary report of K. pneumoniae with MIC susceptibilities had autofiled and was available on the floor terminals at that time. At 5:00 PM on day 3 and order was sent through to discontinue ceftazidime, nafcillin and tobramycin and change the antibiotic order to cefazolin IV 1.0 g q 8h, an appropriate choice for a K. pneumoniae infection and with a hospital cost of $17 per day. The economic impact of this antibiotic modification in response to an interface autofiled result was $123 per day. For a usual 10 day course of therapy the hospital saved $1,230 and the patient was spared the toxicity of an aminoglycoside and a comprehensive assault on her normal flora.

It is apparent that antibiotic usage represents a significant percentage of direct expenditures in the pharmacy budget. This expense is felt more acutely for those inpatients who are under DRG prospective payment since the hospital is reimbursed per diagnosis and not per antibiotic. However, expenditures are converted into savings by clinical decisions to modify antimicrobial therapy in a cost effective manner as shown in Table 3. When an antibiotic is discontinued a direct per diem savings can be demonstrated. During the blood culture audit the date and time of an antibiotic stop order was recorded and the daily savings from that decision was correlated to the reporting of a culture result. Antibiotics that were discontinued following a gram stain report accounted for an average daily savings of $91, and following an interface autofiled report an average savings of $154 per day. Additional savings of $66 per day were recorded for stop orders that followed a printed microbiology report on the patient's chart.

AUDIT OF RESP, WOUND AND URINE CULTURES

After the data from the blood culture audit was reviewed, additional positive cultures from other specimen sources were monitored for the

Table 4.   Resp, Wound, Urine Culture Audit

A.	Adequate Empiric Therapy	12/48 = 25%
B.	Modification Based On:	
	Gram Stain	4/48 = 8%
	Interface Autofiling	8/48 = 17%
	Printed Report	12/48 = 25%
	Report + 1 Day	2/48 = 4%
C.	Inappropriate Therapy	10/48 = 21%

impact of computerized rapid reporting. A total of 48 additional positive patient reports were reviewed which included 14 respiratory, 14 wound and 20 urine cultures. The audit results shown in Table 4 demonstrated that 12 patients (25%) had been placed on adequate empiric therapy. Antibiotic modifications were recorded in 4 patients (8%) following a reported gram stain result and in 8 patients (17%) in response to the on-line autofiled results. Additional modifications were made in 12 patients (25%) following the charting of a printed report and in 2 patients (4%) after the report had been charted for 1 day. Inappropriate therapy or no change in medication orders were recorded in 10 patients (21%). Fewer patients were placed on empiric antibiotic therapy in this audit than in the blood culture audit which was probably related to severity of infection. Furthermore, physicians were more likely to rely on printed reports than computer terminals before modifying antibiotic therapy in respiratory, wound and urinary tract infections. However, the autofiled reports were responsible for 17% of the patients receiving antibiotic modifications. Most of the patients in the inappropriate therapy/no change category had urinary tract infections that were not treated after 2 days of charted results.

An example of the economic impact of an autofiled result for a urinary tract infection is shown in Table 5. Patient B was an 82 year old female admitted to the intermediate ICU on 3/7/88 with a diagnosis of dehydration and sepsis. A clean voided urine specimen was collected at 2:07 PM on 3/7/88, and at 4:30 PM she was started empirically on ampicillin and gentamicin at a hospital cost of $39/day. On the following morning (3/8) the Vitek GNI/GNS-J card pair was set up from isolated colonies. The cards finalized at 2:30 PM on 3/8 and the report was then available through interface autofiling. The report showed that an E.coli resistant to ampicillin was the causative agent of the UTI. At 5:20 PM on 3/8 the ampicillin and gentamicin were discontinued and the patient was placed on cefuroxime to which the organism was susceptible. The clinical impact was the ability to treat the patient with appropriate therapy on the day after specimen receipt. While cefuroxime was not the least expensive alternative therapy the economic impact was a savings of $15/day which extended to a $90 savings over a 6 day course of therapy.

Table 5.   Patient B

3/7/88		Adm. date, 82 F, dehydration/sepsis	
3/7	1407	urine culture collected	
	1730	start amp. IV 2g q 6h	$19/day
		gent IV 60mg q 8h	$20/day
3/8	1430	autofiled Vitek GNI/GNS	
		> 10^5 /ml E. coli amp R	
	1720	stop amp & gent	$39/day
		start cefurox IV 750 mg q 8h	$24/day
Savings $39-24 = $15/day x 6 days = $90			

Table 6. Combined Positive Culture Audit

Antibiotic Decision	Patients			$Impact*
A. Adequate Empiric Therapy	32/84	=	38%	
B Modification Based On:				
Gram Stain	12/84	=	14%	$ 91/day
Interface Autofiling	18/84	=	21%	243/day
Printed Report	16/84	=	19%	229/day
Report + 1 Day	4/84	=	5%	0
C. Inappropriate Therapy	12/84	=	14%	

*Average daily savings due to antibiotic stop orders.

Although the audit of respiratory, wound and urine culture indicated that antibiotic modifications occurred in 4 patients following receipt of gram stain results, none of the modifications resulted in an antibiotic stop order and thus no direct daily savings for the pharmacy could be documented. However, these patients were started on appropriate therapy on day 2 which was an indirect contribution towards reducing length-of-stay. Antibiotic stop orders placed in response to reports autofiled by the Vitek-Sunquest interface resulted in an average daily savings of $89, and in response to printed reports the savings averaged $163/day.

COMBINED DATA OF POSITIVE CULTURES

The combined positive culture results shown in Table 6 indicated that adequate empiric antimicrobial therapy was documented in 32 of the 84 total patients (38%). Antibiotic modifications occurred in the treatment of 12 patients (14%) following a reported gram stain result, and accounted for an average savings to the pharmacy of $91 per day. Changes in therapy following the Vitek-Sunquest interface autofiled reports were documented in 18 patients (21%) and the direct economic impact as measured by antibiotic stop orders averaged $243 per day. Antibiotic changes based on charted reports occurred in 16 patients (19%) with a average savings of $229 per day. As a general rule the interface auto-filed reports were most likely to have a rapid impact on patient management and antimicrobial therapy if the report was associated with a patient on the infectious disease service, trauma service or any of the teaching services where residents and medical students interacted frequently with computer terminals.

CONCLUSIONS OF AUTOFILING STUDY

1. Adequate empiric antimicrobial therapy was documented in 38% of patients with positive culture results.

2. Modifications in antibiotic therapy due to rapid automated reporting of bacterial identification and susceptibility results were documented in 21% of patients with positive culture results.

3. The direct economic impact of the Vitek-Sunquest interface autofiling function accounted for an average savings of $243 per day to the pharmacy.

4. Rapid microbiology reporting provided the ability to make earlier clinical decisions, indirectly contributing towards reducing the patient's length-of-stay.

BIDIRECTIONAL COMPUTER INTERFACE IMPLEMENTATION

An additional software enhancement for the Vitek-Sunquest interface was the development of a Bidirectional Computer Interface (BCI). The BCI allows two-way communication between the Vitek Information Management System (IMS) and a laboratory host computer. The new BCI software not only improves the result autofiling function but also expands reporting capabilities by using the IMS program to generate customized patient reports and a wide variety of epidemiology reports. Transmission of data using the BCI occurs through separate upload and download functions. The upload function automatically transfers completed test results from the Vitek computer to the host computer (Sunquest). The user may also manually upload test results through the Vitek keyboard using the "bcisend" command. The download function transfers patient, specimen and culture demographics from the host computer to the Vitek computer which occurs by manual entry through the host Sunquest keyboard. Demographics may be downloaded at any time by entering the function code "BCI" followed by the selected culture accession numbers. The patient, specimen and culture information that are linked to an accession number are downloaded and filed in the Vitek IMS by the patient identification number. In this manner multiple specimen/culture demographics from the host computer can be filed under a single patient identification number in the Vitek IMS. If both the upload and download functions are transmitting at the same time the host computer always has priority and therefore download will transmit first followed by upload.

PREREQUISITES FOR BCI

To be compatible with the BCI software the user Vitek System must have the DSIMS-R2.01 or a later version software and must be equipped with a 50 or 80 megabite (Mb) hard disk, 1Mb of main memory and at least one available RS-232C serial port. For Sunquest compatibility our Vitek serial port characteristics were set at a 1200 baud rate, even parity, 7 bits per character and 1 stop bit. An essential prerequisite for the download sequence was the establishment of the same demographic codes for both Sunquest and Vitek.

The existing host computer Sunquest codes which were all uppercase or numerals were defined in the same manner in the Vitek IMS user table maintenance for four user tables. Sunquest location (LOC) codes were (PHYS) codes which were four numerals were defined the same way in the Vitek physician (pp) table. Likewise, the codes for Sunquest specimen type (SDES) and culture battery (TEST) were defined in the Vitek tables of specimen source (ss) and culture type (ct) respectively.

BCI MESSAGE TRANSMISSION

The BCI software has been designed in two separate levels: a communications level and an applications level. The communications level maintains the structural sequence of an information exchange between the host computer and Vitek, whereas the applications level defines the actual content of the messages themselves. A download sequence from the communications level is illustrated in Table 7 by a sequence of character signals. The session or information exchange is initiated by the host Sunquest computer transmitting the character 'ENQ' which is a line bid character. The Vitek then responds with an acknowledge character 'ACK', The actual beginning of a message occurs with the host sending a start of text character 'STX'. The record separator characters 'RS' then follow which are the beginning characters for a data stream defined in the applications layer. For example the communications level 'RS' character

Table 7. DOWNLOAD SEQUENCE

```
SUNQUEST (HOST)                    VITEK (BCI)
        ENQ ------------------->
        <-------------------ACK
        STX ------------------->
        RS (Data 1) ----------->
        RS (Data 2) ----------->
        RS (Data 3) ----------->
        GS (checksum) --------->
        <-------------------ACK
        ETX ------------------->
        EOT ------------------->
```

would be followed by an applications level group of characters (Data 1) including patient name, patient ID, date of birth and sex. The next 'RS' would be followed by patient location, hosp. service, admitting diagnosis, primary physician, adm. date. After all of the records have been transmitted a group separator character "GS" is sent which carries with it a checksum function to detect transmission errors. The Vitek would then respond with an 'ACK' if all data were received correctly or the Vitek could respond with a negative acknowledge character 'NAK' if there were a problem with the BCI transmission. In the case of a 'NAK' the host would retransmit the entire packet of data associated with that particular checksum i.e. RS (Data 1), RS (Data 2), etc. When the 'ACK' is received by the host, an end of text character 'ETX' is sent by the host which signals the end of the communications packet. Finally an end of transmission character 'EOT' is sent by the host to signal the end of the session. An upload sequence would follow the same communications level format in the opposite direction with the Vitek initiating the session with 'ENQ.'

BCI APPLICATIONS LEVEL

The applications level of the BCI contains the actual descriptions of the transmitted message. It is at this level that the true flexibility and customization of the BCI can be appreciated because the user selects the field names associated with patient, specimen and culture demographics. The utilization of patient fields which are shown in Table 8 depends on the defined codes in the Vitek IMS user tables and the format of the IMS [log] screen for "Enter Patient Information." Each field name in Table 8 also has a corresponding maximum length, but this too can be altered to fit within the data base. The minimum requirement for downloading patient demographics is the presence of a patient ID number which is provided by the host computer.

Table 8.   DOWNLOADED PATIENT FIELD TYPES

Field	Description	Length
pn	Patient Name	30
pi	Patient ID	12
pb	Patient Date-of-Birth	8
ps	Patient Sex	10
pl	Patient Location	6
px	Patient Hospital Service	6
po	Patient Adm. Diagnosis	6
pp	Patient Primary Phys.	6
pa	Patient Adm. Date	8

TABLE 9.   DOWNLOADED SPECIMEN/CULTURE FIELD TYPES

Field	Description	Length
si	Specimen ID	4
ss	Specimen Source	6
st	Specimen Site	6
sl	Specimen Location	6
sp	Specimen Request Phys.	6
sx	Specimen Service	6
s1	Spec. Collection Date	8
s2	Spec. Collection Time	5
s3	Spec. Receipt Date	8
s4	Spec. Receipt Time	5
sc	Specimen Comment	10
ci	Culture ID (Exam ID)	10
ct	Culture Type	6

Any combination of specimen fields listed in Table 9 can be downloaded and filed with patient demographics. Specimen fields that have not been included in the IMS [log] screen and fields that are not completed in the host computer will be ignored by the Vitek. Specimen demographics that are downloaded will be provided with a consecutive specimen ID by the Vitek relative to the total number of specimens filed for a given patient, i.e. specimen 1 of 1, 1 of 2, etc. The minimum requirement for downloading specimen demographics is the presence of a culture ID and culture type for each specimen. The culture ID is the accession number assigned by the host computer and corresponds to the number on the Vitek cards for that specimen, sometimes referred to as the examination ID. The culture type is defined by the host computer as routine, fungal, AFB or more specifically, blood culture, respiratory culture, urine culture, etc. The association of one patient ID with multiple culture ID's by both host and Vitek allows many specimen/culture demographics to be filed with one patient ID and one set of patient demographics in the IMS.

BCI DOWNLOAD SEQUENCE

The actual download sequence for the Sunquest host computer begins with the entering of the function code 'BCI'. An accession number prompt then appears and selected culture accession numbers are entered as shown in Table 10. The Sunquest accession numbers start with an alpha character representing the day of the week followed by up to four numbers. The Sunquest portion of the BCI converts the accession number to a five digit number for compatibility with Vitek with each alpha day of the week converted to a numerical day, 1 through 7.

TABLE 10.   SUNQUEST DOWNLOAD
FUNCTION: BCI

MANUAL MICROBIOLOGY LOADLIST
FOR VITEK

ACC NO: T5445
ACC NO: T5657
ACC NO: T5807
ACC NO: W735
ACC NO: W886
ACC NO: <CR>

TABLE 11.   SUNQUEST DOWNLOAD

MANUAL MICROBIOLOGY LOADLIST FOR VITEK

ACC NO	PATIENT NAME	PAT NO	LOC	CUL	SOR	COLLECT DATE
T5445	CAMDEN, JOSEPH	0222097	A8	BCL	BLUD	04/26/88
T5657	COOPER, CATHERINE	04006293	IICU	RESP	SPU	04/26/88
T5807	TRUMAN, HARRY	00541942	N10	WNDS	LSN	04/26/88
W735	DULLAS, DEBBIE	04004821	0-A	URC	UCV	04/27/88
W886	ADAMS, SAMUEL	02126107	N8	URC	UCAT	04/27/88

ACCEPT (A), MODIFY (M) OR REJECT (R)? A

The Sunquest terminal then displays the loadlist for Vitek as shown in Table 11.   The screen displays the chronological order of accession numbers followed by patient name and patient identification number.   The next three fields of location, culture type and specimen source are codes that have been identically defined in the IMS user tables.   The last field on the screen is the collection date.   By accepting this loadlist the Sunquest proceeds to download each record and indicates on the screen that the record has been 'accepted.'

VITEK IMS SCREEN

To take advantage of the numerous patient and specimen fields that could be downloaded, the long log format of the IMS was selected (Table 12).   However, the long log was modified and rearranged to include patient, specimen and culture fields on one screen. By selecting the IMS option 'Enter Patient Information' and upon entering the patient ID or culture ID number the downloaded information is displayed. The download portion of the BCI thus saves considerable personnel time for the keyboard entry of demographics. Moreover, the user now has access to a variety of organism and susceptibility reports with the flexible sorting capabilities of the Vitek IMS.

TABLE 12. [log 1] VITEK IMS SCREEN

```
Patient Id: 04006293    Patient Name: COOPER, CATHERINE
Patient Location: IICU  INTERMEDIATE ICU
Date of Birth: 02/04/21   Sex:  F
Admission Date:  04/06/88  Dosage Level Group:  A Adult
Attending Physician: 1106  SNYDER, HARVEY:  MED
-----------------------------------------------------------
Specimen Source: SPUT  SPUTUM   Specimen No: 1 of 1
Specimen Site:
Collection Date:  04/26/88  Collection Time:  06:15
Receipt Date:  04/26/88    Receipt Time:  07:30
Requesting Physician: 1149   ALESSI, PAUL:  MED
Specimen Nocation:  N7  NORTH 7
-----------------------------------------------------------
Culture ID No:  35657                Exam No.:  1 of 1
Culture Type:  RESP    RESPIRATORY CULTURE
Comments       of
Final Date:    Final Time
```

SUMMARY

The rapid automated reporting capability of the Vitek-Sunquest interface was shown to have a direct clinical and monetary impact on patient management. The audit performed on autofiled positive culture results at Cooper Hospital/University Medical Center demonstrated an average savings of $243 per day in antimicrobial utilization. A further enhancement in the computer software for the Vitek and Sunquest systems was the development of the Bidirectional Computer Interface (BCI). The BCI upgrade allows the Microbiology laboratory to utilize both the Sunquest and Vitek IMS programs to generate a wide variety of epidemiology reports with a minimum of keyboard entry time.

ACKNOWLEDGEMENT

The Vitek-Sunquest Bidirectional Computer Interface represents the cooperative efforts of individuals at three institutions. At Vitek Systems I thank Betsy Fortman, Rod Landers, Ross Livengood, Joe Wiegner, Mark Lovell, Barb Shannahan, Cathy Wensauer and Mike Moore for the Vitek BCI development and the integrated support for its implementation. At Sunquest Information Systems I thank Audrey Carlson, Anita Hollsinger, Mike Medachy and Paul Syml for the Sunquest BCI development. And at Cooper Hospital I thank Lina DeSantis for her computer assistance and Eleanor Hofbauer for her secretarial assistance.

Pure culture technique, 1
Putrefaction, 1, 2
PYR (L-pyrrolindonyl-β-naphthylamide), 38
PYR hydrolysis, 41

Radioactive sodium iodide, 80
Radioimmunoassay (RIA), 80, 125
Recurrent sepsis, 135
Respiratory syncytial virus (RSV), 115, 117, 118, 125
Restrictive endonucleases, 98, 101
Rheumatic fever, 37, 43-45
Rhodococcus equi, 52
R. rubropertinctus, 52
R. terrae, 52
Ribonuclease, 26
RNA:DNA hybridization, 25, 26
Ribosomes, 22
rRNA, 22, 23, 26
Rotavirus, 115, 116, 125, 126
Rubella, 126

Salmonella, 18
Samarium (Sm), 124
Sepsis, 136
Serologic tests, 77, 126
Sheep's blood agar, 38
Shell vial assay, 118
Sickle cell anemia, 17
Simian virus 40 (SV 40), 92
Single-photon counting fluorometer, 124
Sodium bisulfite, 127
Solution hybridization, 129
Southern blot analysis, 90, 99-102
Specimen/culture demographics, 140
Specimen toxicity, 119
Spontaneous generation, 1
Staphylococci, 133
Stokes shift, 124
Streptavidin-horseradish peroxidase, 73

Streptococcal antigen detection, 38, 40
 impact on patient management, 43
Streptococcal infection, 37
Streptococcus bovis, 111
S. pneumoniae, 112
Stringency, 98
Sulfonated probe, 61
Sunquest information systems, 132, 134, 141
Syva test, 113

Teichoic acids, 4
Terbium (Tb), 124
Test-of-cure evaluation, 86
Tetracycline, 113
Time-resolved fluorometry, 123-129
Tonsillar/retropharyngeal abscess, 37
Trans-acting transcriptional enhancer, 95
Trans-acting transcriptional repressor molecule, 95
Tuberculosis complex probe, 28, 29
Turnaround time (TAT), 131, 132

Urine antigen detection, 58, 67

Vancomycin-containing medium, 84
Vesicular stomatitis, 29
Varicella-zoster virus (VZV), 117
Vibrio cholerae, 2
Viral antigens, 123
Viral capsid antigen, 94
Viral infections, 115-119
Viral isolation, 116
Viral mRNA transcripts, 103
Viral transport medium (VTM), 72
Virions, 89
Vitek IMS, 138, 140-142
Vitek-Sunquest interface, 132, 137, 142
Vitek systems, 132, 141